International Law and Renewable Energy Investment in the Global South

This book will discuss the legal tools offered by international law that can support foreign direct investment (FDI) in the renewable energy sector in the Global South.

Promoting and increasing investment in the renewable energy sector is crucial for limiting global temperature rise to 1.5°C and addressing energy poverty in the Global South. In this volume, Avidan Kent explores the various home-country measures (HCMs) offered by international law that support FDI in the renewable energy sector. This book provides a bird's eye evaluation of HCMs from fields such as trade law, investment law, environmental law, development law and more. It reveals that while international law indeed offers many legal tools to support investors' needs, the current legal framework is fragmented; most legal instruments were designed in isolation and the potential for mutually supportive, synergetic policies has been explored only to a limited extent. This fragmented reality is in contradiction to the notion of Policy Coherence for Development, which is increasingly gaining support in leading institutions in Europe and elsewhere. This book will provide recommendations on the manner in which HCMs can be connected in order to maximise their potential and boost investment in renewable energies in the developing world.

International Law and Renewable Energy Investment in the Global South will be of great interest to scholars, students and practitioners of international law, energy studies, development studies and IR more broadly.

Avidan Kent (Cantab) is an Associate Professor at the University of East Anglia, UK.

Routledge Explorations in Energy Studies

Electricity and Energy Transition in Nigeria
Norbert Edomah

Renewable Energy Uptake in Urban Latin America
Sustainable Technology in Mexico and Brazil
Alexandra Mallett

Energy Cooperation in South Asia
Utilizing Natural Resources for Peace and Sustainable Development
Mirza Sadaqat Huda

Perspectives on Energy Poverty in Post-Communist Europe
Edited by George Jiglau, Anca Sinea, Ute Dubois and Philipp Biermann

Dilemmas of Energy Transitions in the Global South
Balancing Urgency and Justice
Edited by Ankit Kumar, Johanna Höffken and Auke Pols

Assembling Petroleum Production and Climate Change in Ecuador and Norway
Elisabeth Marta Tómmerbakk

International Law and Renewable Energy Investment in the Global South
Avidan Kent

For more information about this series, please visit: www.routledge.com/Routledge-Explorations-in-Energy-Studies/book-series/REENS

International Law and Renewable Energy Investment in the Global South

Avidan Kent

Routledge
Taylor & Francis Group
LONDON AND NEW YORK

earthscan
from Routledge

First published 2022
by Routledge
2 Park Square, Milton Park, Abingdon, Oxon OX14 4RN

and by Routledge
605 Third Avenue, New York, NY 10158

Routledge is an imprint of the Taylor & Francis Group, an informa business

British Library Cataloguing-in-Publication Data
A catalogue record for this book is available from the British Library

Library of Congress Cataloging-in-Publication Data
Names: Kent, Avidan, author.
Title: International law and renewable energy investment in the global south / Avidan Kent.
Description: Milton Park, Abingdon, Oxon ; New York, NY : Routledge, 2021. |
Series: Routledge explorations in energy studies | Includes bibliographical references and index.
Identifiers: LCCN 2021015447 (print) | LCCN 2021015448 (ebook)
Subjects: LCSH: Investments, Foreign (International law) | Insurance law. | Foreign trade regulation. | Environmental law, International. | Renewable energy sources—Law and legislation.
Classification: LCC K3830 .K46 2021 (print) | LCC K3830 (ebook) | DDC 346/.092—dc23
LC record available at https://lccn.loc.gov/2021015447
LC ebook record available at https://lccn.loc.gov/2021015448

ISBN: 978-0-367-22858-3 (hbk)
ISBN: 978-1-032-10076-0 (pbk)
ISBN: 978-0-429-27720-7 (ebk)

DOI: 10.4324/9780429277207

Typeset in Times New Roman
by codeMantra

For Joëlle
For Arielle
For Auri
For Rakefet

Contents

Acknowledgements

I wish to thank Sophie Kelley and Holly Hancock for their excellent research assistance. I am grateful to Annabelle Harris and Matthew Shobbrook (Routledge) for their support (and patience) throughout this process.

I wish to express my gratitude to Michael and Chana Kent (my parents) for their support.

List of acronyms

ACP	African, Caribbean and Pacific Group of States
ADB	Asian Development Bank
APEC	Asia-Pacific Economic Cooperation
ASEAN	Association of Southeast Asian Nations
ATIA	African Trade Insurance Agency
BIT	Bilateral Investment Treaty
CACM	Central American Common Market
CAO	Office of Compliance Advisor/Ombudsman
CDM	Clean Development Mechanism
COMESA	Common Market for Eastern and Southern Africa
COP	Conference of Parties
CPI	Climate Policy Integration
CSR	Corporate Social Responsibility
CTS SS	Committee on Trade and Services Special Session
DFC	Development Finance Corporation
EBRD	European Bank for Reconstruction and Development
ECOWAS	Economic Community of West African States
ECT	Energy Charter Treaty
EFTA	European Free Trade Association
EGA	Environmental Goods Agreement
EGS	Environmental Goods and Services
EIA	Environmental Impact Assessment
EIB	European Investment Bank
EPA	Economic Partnership Agreement
EPI	Environmental Policy Integration
ESA	Eastern and Southern Africa
ESIA	Environmental and Social Impact Assessment
FDI	Foreign Direct Investment
FET	Fair and Equitable Treatment
FIT	Feed-in Tariffs
FTA	Free Trade Agreement
GATS	General Agreement on Trade in Services
GATT	General Agreement on Tariffs and Trade

GCF Green Climate Fund
GEF Global Environment Facility
GHG Greenhouse Gases
HCM Home Country Measures
HRIA Human Rights Impact Assessment
ICSID International Centre for Settlement of Investment Disputes
IDB Inter-American Development Bank
IEA International Energy Agency
IFC International Finance CorporationIIA International Investment Agreement
IISD International Institute for Sustainable Development
ILO International Labour Organization
IP Intellectual Property
IPCC Intergovernmental Panel on Climate Change
IRENA International Renewable Energy Agency
IRM Independent Redress Mechanism
ISDS Investor-State Dispute Settlement
IUCN International Union for the Conservation of Nature
KP Kyoto Protocol
LCR Local Content Requirement
LDC Least Developed Countries
MDB Multilateral Development Bank
MIGA Multilateral Investment Guarantee Agency
NDC Nationally Determined Contributions
ODA Official Development Assistance
OPIC Overseas Private Investment Corporation
PCD Policy Coherence for Development
PCSD Policy Coherence for Sustainable Development
PPM Parts Per Million
PR Performance Requirements
PRI Political Risk Insurance
PTA Preferential Trade Arrangement
PV Photovoltaic
RE Renewable Energy
SADC Southern African Development Community
SCM Agreement on Subsidies and Countervailing Measures
SDG Sustainable Development Goals
SDM Sustainable Development Mechanism
SETA Sustainable Energy Trade Agreement
SRMI Solar Risk Mitigation Initiative
TiSA Trade in Services Agreement
TNC Transnational Corporations
TRIPS Agreement on Trade-Related Aspects of Intellectual Property Rights
UEMOA West African Economic and Monetary Union

UKEF	UK Export Finance
UNCTAD	United Nations Conference on Trade and Development
UNEP	UN Environment Program
UNFCCC	United Nations Framework Convention on Climate Change
UNGP	United Nations Guiding Principles on Business and Human Rights
USMCA	Agreement between United States of America, Mexico and Canada
WIR	World Investment Report
WTO	World Trade Organization

Introduction

> We, the Heads of State and Government and High Representatives, meeting at United Nations Headquarters in New York from 25 to 27 September 2015 as the Organization celebrates its seventieth anniversary, have decided today on new global Sustainable Development Goals. [...]
>
> - Ensure access to affordable, reliable, sustainable and modern energy for all [...][SDG 7]
> - Take urgent action to combat climate change and its impacts [...][SDG 13]
>
> We commit ourselves to working tirelessly for the full implementation of this Agenda by 2030.[1]
>
> The Sustainable Development Goals

I The twin challenge: access to energy and climate change

With the above words, the international community committed itself to urgently and effectively addressing energy poverty around the world. A significant part of this effort will have to take place in the Global South, where access to energy is most needed and investment conditions are uninviting. While the exact figures and timescales will be presented in Chapter 1 of this study, it is suffice to say, at this early point, that they are staggering. More than 1 billion individuals currently live without access to energy. This energy poverty is leading to a myriad of subsequent problems, which range from slow economic development to poor health and educational services, inequalities, poverty and hunger.

As if the task of addressing energy poverty in the Global South is not challenging enough on its own, the Sustainable Development Goals (SDGs) demand that this effort be achieved in a sustainable manner. Notably, SDG 13 demands an *urgent* and *aggressive* climate action. Simply increasing access to energy through cheap, fast and polluting solutions will not do; it will lead to a head-on collision between SDG 7 (access to energy) and SDG 13

1 United Nations General Assembly, *Transforming Our World: The 2030 Agenda for Sustainable Development* (2015) A/RES/70/1.

DOI: 10.4324/9780429277207-1

(climate action), solving one problem by enhancing another. This is a trade-off that countries in the Global South should not be tempted to make: the economic and societal impacts of climate change will be felt most acutely in these regions, and are expected to be catastrophic in both nature and scale.

The imperative to reconcile climate action and energy poverty points to one obvious solution: the significant dissemination of clean renewable energies (RE) in the Global South. As explained in Chapters 1 and 2 of this study, this increase will only take place with the extensive involvement of those who possess the capital, knowledge and technology required for such a momentous task, namely, private sector investors.[2] While private foreign direct investment levels are certainly increasing, this process is not happening fast enough; not for the planet and certainly not for those who are living without access to energy. Far more private investment is necessary, and at a much greater pace.[3]

The task of increasing private investment in the RE sector in the Global South is nothing short of monumental. As explained in Chapter 1 of this study, a variety of investment barriers inhibit international efforts to increase investment in the RE sector and prevent the free market from doing what it does best: providing a product where ample demand exists. Addressing these barriers and achieving these goals will require a multitude of solutions; most of them will have to be supported through policies and laws,[4] which in some cases will have to be *international* in nature, implying a role for *international law.*

II The need for policy coherence

Chapters 3–7 of this study will address the role that international law can play in meeting the above-described challenge. The topics addressed in these chapters will vary to a great extent: ranging from international trade rules on areas such as goods and services, subsidies and intellectual property, to investment protection and promotion rules, insurance mechanisms, environmental laws, human rights safeguards, development treaties, development institutions and more.

This lack of focus is intentional and planned: these areas are all important in the effort to increase investment in the RE sector in the Global South, and each of them addresses different investment barriers and challenges. Private investors consider elements such as the cost of finance, trade liberalisation, the ease of doing business, the state of existing infrastructure (including legal infrastructure), the safety and stability of the business environment and sufficient guarantees. Addressing these aspects will require a *comprehensive*

2 The estimates of required investment are discussed in Chapter 1 of this study.

3 See review in Chapter 1 of this study.

4 OECD, *Aligning Development Co-operation and Climate Action: The Only Way Forward* (OECD 2019), 22.

legal answer, one that will necessitate a variety of legal reforms. The challenge, so it seems, is multifaceted, as will need to be the solution.

Moreover, the variety of issues discussed in these chapters are all inter-linked and inter-dependent. Addressing one area will support the effectiveness of another. For example, providing investment guarantees and insurance mechanisms will reduce the cost of finance. Equally, reducing trade barriers will increase the effectiveness of Official Development Assistance (ODA), and securing environmental and human rights safeguards will ensure the sustainability and durability of RE projects. The focus of this research, therefore, cannot be only investment law, or only development law: the investment environment is far richer and much more complex than that.

One example that explains the necessity of addressing this myriad of topics holistically[5] is the reliance on ODA as a tool for supporting the transition into a green economy. It is widely accepted today that on their own, ODA funds will not be enough to address this challenge. To begin with, the relevant sums are insufficient: current public transfers, as they stand, are far from what is required to finance the global transition into a green economy.[6] As discussed elsewhere in this book, developed nations are aware of this deficiency and increasingly view ODA funds *as a leveraging tool*, that will, hopefully, attract much bigger sums from the private sector ('from billions to trillions'). To date, however, this effort has not been successful and the leveraging power of ODA funds remains low.[7] As argued in this study, other policies, from a myriad of fields, will have to fill this leveraging gap and improve ODA's value for money.

This background implies a comprehensive and complex challenge ahead: addressing energy poverty in the Global South will require more than ambitious financial contributions. It will require a significant overhaul of the international legal governance structure, and, importantly, more attention to *policy coherence*. Addressing one investment barrier while ignoring others will not achieve optimal results, and, as stated above, optimal results are what the urgency of the situation at hand currently requires. This conclusion is not novel on its own. Indeed, the Addis Ababa Action Agenda on Financing for Development, which was concluded in order to support the implementation of the SDGs, addresses the challenge of policy coherence[8]:

> 'We will respect each country's policy space and leadership to implement policies for poverty eradication and sustainable development, while

5 Many other examples are reviewed in Chapters 3–7.

6 See more on this topic in Chapter 1 of this study.

7 Samantha Attridge and Lars Engen, *Blended Finance in the Poorest Countries* (Overseas Development Institute 2019) 11 <https://www.odi.org/sites/odi.org.uk/files/resource-documents/12666.pdf>.

8 Addis Ababa Action Agenda of the Third International Conference on Financing for Development (UN 2015), para 9. Emphasis added.

> remaining consistent with relevant international rules and commitments. At the same time, national development efforts need to be supported by an enabling international economic environment, including coherent and mutually supporting world trade, monetary and financial systems, and strengthened and enhanced global economic governance. [...]' 'We commit to pursuing policy coherence and an enabling environment for sustainable development at all levels and by all actors, and to reinvigorating the global partnership for sustainable development'.

The fact that so many different areas of law (and areas of *expertise*) are relevant for the design of coherent laws and policies imposes major difficulties for policy-makers. This struggle is not new on its own: it is widely addressed in International Relations literature on institutional interactions and academic legal literature on the fragmentation of international law. As reviewed throughout this book, in the context of this study – RE investment in the Global South – policy-makers have vastly failed to address this problem. It is clear that the intentions are there; there is no shortage of declarations and statements that demonstrate how the international community is aware of this challenge and the need to improve laws and policies by increasing coherence. What is still missing are the next steps forward: the *concretisation* of this effort through the identification of specific synergies and the design of concrete, coherent laws and policies that will support the attainment of SDG 7.

As reviewed in the next part, this study will take one, modest step in this direction.

III The study's scope, focus and limits

This book addresses the challenge of increasing investment in the RE sector in the Global South. It critically reviews the efforts that have been made in several areas of international law in order to address the existing investment gap. It identifies gaps in the regulations, and, importantly, comments on the lack of coherence in existing international legal frameworks.

As mentioned above, the task faced by the international community is colossal in scope. This study does not address this challenge in its entirety. The focus of this study will therefore be narrowed on three levels: (1) the type of laws that will be discussed (i.e. *international* law), (2) the type of measures that will be addressed (i.e. *Home Country Measures*) and (3) the type of policy design that will be promoted (i.e. Policy Coherence for Development). Below is a short review of each of these elements, explaining the decision to address each of them in this study.

A International law

First, this is a study of *international law*. As explained in Chapters 1 and 2 of this book, addressing the investment gap in the RE sector in the Global

South will require a myriad of policies, both *international* and *domestic* in nature. However, the author of this study is an international law scholar, so will therefore focus almost exclusively on international laws and institutions. Beyond the author's expertise, the focus on international law is also justified for other reasons: the global nature of the challenges at hand; the importance of international cooperation in addressing these challenges (notably the need to engage the resources of *developed* nations with the needs of *developing* countries) and the role that international law plays in improving investment conditions – all of these issues (and others as well, as discussed in Chapter 2) provide sufficient reason to justify the focus on international law in the context of this study.

Importantly, the focus on international law does not imply that domestic laws and policies are not vital – they certainly are. Domestic regulatory reforms are at least as important as international laws and policies and will also have to be addressed. However, the sheer scope of the challenge at hand requires certain delimitations and, for now, domestic laws and policies will need to be addressed by others, preferably by experts in specific domestic legal systems.

B Home country measures (HCMs)

Second, the international law universe is vast. Accordingly, this study will not cover *all* relevant international laws. It will primarily address a specific type of international law: *international law Home Country Measures* (*HCMs*), a term that will be explained and defined in Chapter 2 of this study. As explained in Chapter 2, the focus on HCMs is justified for several reasons. First, the investment gap will not be solved unless active steps are taken by Global North nations, where the necessary technologies and resources exist. Second, and as is widely recognised in almost every international climate change agreement or decision,[9] the Global North carries most of the responsibility for addressing the problems at hand. Third, some of the necessary key measures are in effect controlled, determined, or at the least highly affected by the Global North (e.g. ODA transfers, or investment promotion activities). These three reasons imply a significant, and much more active role for developed nations, i.e. investors' *Home Countries*. This role should be filled with content: notably, developed nations should commit to ambitious HCMs, i.e. international laws that actively increase outwards RE-related investment in the Global South. These efforts cannot be left to developing nations' attempts to *attract* investment (i.e. Host Country Measures), as the responsibility, as stated above, lies elsewhere.

9 Almost all UNFCCC decisions and agreements are stressing the importance of the principle of common but differentiated responsibilities, and the implied role that developed nations must play in order to implement this principle.

Another justification for the focus on international law HCMs is the existing gap in the literature. Simply put, literature on *Host* Countries Measures is abundant; mountains of books, articles and dissertations have been written about developing nations' interaction with foreign investors. On the other hand, the universe of HCMs is defined as 'unexplored'.[10] Therefore, the focus on HCMs seems more useful as it will address a well-recognised gap in the literature.

C Policy coherence for development (PCD)

Finally, this study will address the levels of *policy coherence* currently found in international law HCMs. As explained in Chapter 2 of this study, the examination of policy coherence in the context of this study (i.e. with a focus on the Global South) must be conducted through the more specific lens of the term Policy Coherence for Development. Importantly, this term frames the discussion on policy coherence within the context of development, setting clearer development-related objectives at the core of the attempts to improve coherence and mandating a specific type of mainstreaming.

More concretely, when reviewing the different HCMs, this study will assess whether these were designed in isolation, or whether attempts to link them with other HCMs were made in order to improve their effectiveness. Policy coherence, it will be explained, is crucial for improving the effectiveness of policies that are aimed at increasing investment in the RE sector in the Global South.

IV The study's structure

Bearing these three themes in mind, this study will take the following shape. Chapter 1 of this study will set the scene and answer three preliminary questions: the 'Why?', the 'Who?' and the 'What?'. The 'Why?' question addresses the study's *raison d'être*; it explains the need for public intervention (i.e. laws and policies) in the context of SDG 7. The 'Who?' question addresses the need to target private foreign direct investors as the subjects of laws and policies, and explains the role and importance of these actors in the context of this study. Finally, the 'What?' question asks what the objective of these laws and policies should be? It addresses the importance of integration in terms of both means and outcomes.

Chapter 2 of this study will then address the 'How?' question, or, in other words, the tools that are required for the attainment of the grand objective of increasing investment in the RE sector in the Global South. It introduces,

10 Karl Sauvant et al. 'Trends in FDI, home country measures and competitive neutrality' in Andrea Bjorklund (ed.) *Yearbook on International Investment Law & Policy 2012–2013* (OUP 2014) 3 <http://ccsi.columbia.edu/files/2014/03/Yb-12-13-ch.-1-8-Nov-13-stand-alone-final-for-website.pdf>.

defines and discusses the role of two key concepts: international law Home Country Measures ('international law HCMs') and Policy Coherence for Development (PCD). These two concepts – international law HCMs and PCD – will guide the rest of this study and will be at the heart of the assessment conducted in Chapters 3–7.

Chapters 3–7 will assess specific international law HCMs, and evaluate the presence of PCD in their design. More specifically, these chapters will address the following fields of law: international investment law, international trade law, international development law, international insurance law, international environmental law and human rights law. The chapters will ask whether these international law HCMs were designed in isolation, and what type of linkages should be enforced. While this study does not present a comprehensive policy proposal (e.g. in the shape of a model treaty, or model laws), it does present certain recommendations that policy-makers may wish to consider.

V And what about...?

This study provides an extensive review of the different international laws that are relevant for the promotion of RE-related investment in the Global South. However, importantly this book is not a textbook, and it does not cover *each and every* angle that could be of relevance for the attainment of SDG 7. The expert reader will undoubtedly identify additional areas that, in his/her view, should have been added to this review. Without a doubt, the expert reader will be correct in this assessment.

The author will nevertheless answer this potential criticism by stating that he addressed the areas that, in his view, deserve close scrutiny and that, naturally, a certain process of prioritisation (and also elimination) had to take place. The scope of 100,000 words and two years of labour will have to suffice for one study alone, and the author will invite the expert reader to fill any identified gaps with his/her own analysis.

Moreover, and perhaps less defensively, the author will add that, once again, this is not a textbook. This is a monograph aimed at achieving a goal that is different from the mere review of each and every aspect of the law. To begin with, while this study does provide a comprehensive, technical and legal review, it is also a study of *integration*. It is now clear that efforts made to combat climate change and to achieve SDG 7 must be cross-sectorial and interdisciplinary. Indeed, the international community committed itself to the integrationist SDGs and the 'leave no one behind' moto. The topic of this book represents a useful case study for the evaluation of how integrated in fact states' policies are, and how much coherence indeed exists in relevant international laws. And indeed, the picture portrayed throughout this study clarifies the enormity of the challenge of integration. This study demonstrates that significant attention – and resources – will have to be invested in addressing the issue of policy design. This study will therefore call for reforms not only in law, but also in the law-making process: it will advocate for a more 'bird's-eye view', more academic research of linkages, and far less atomisation in the process of policy-making.

Lastly, this study is also important in shedding light on the significance and the usefulness of international law HCMs. This type of measure is much less explored in academia, and, unfortunately, in some cases, also in the practice of international law. As stated above, there are many reasons to focus on international law HCMs, and states will do well to explore these possibilities more seriously.

As stated in the conclusion of this book, this study should be addressed as a starting point, and not as a final word. The more one researches a topic, the more he discovers the vastness that remains unaddressed. It is hoped that this study will be of use for academics and policy-makers alike, and that it will inspire others to explore this extremely important, yet somewhat unaddressed area of international law.

1 Investment in the renewable energy sector in the Global South

I Introduction: the why, the who and the what

This study examines international efforts to increase investment in the renewable energy (RE) sector in the developing world. The focus of this study is international law: a fragmented and crippled legal system that currently endures a political backlash and a crisis of legitimacy. This system, nevertheless, possesses powerful tools: tools that can, and already do, deliver. Through the assessment of several sub-fields of international law (investment, trade, environment, insurance and development), this study examines several existing (and negotiated) legal instruments. It will evaluate these laws' legal architecture, and their effectiveness in achieving the international efforts to increase RE-related investment in the developing world.

The first chapter of this book will place the foundations for this study. It will begin by explaining the study's parameters, presented as a series of 'why', 'who' and 'what' questions. A fourth question – 'how' – will be addressed in Chapter 2, and throughout the rest of this study.

The first question – 'why' – is asked in order to explain the study's *raison d'être*: why is it necessary to design policies in order to increase investment in the RE sector? Why should states intervene with the operation of the market's free hand? And why is it necessary to focus on investment in the Global South, of all places? In answering these questions, this study will discuss the Sustainable Development Goal (SDG) 7, as well as more specific justifications such as market barriers and incentives that could be addressed only through governments' interventions.

Next, this chapter will address the 'who' question: who should the policies to increase investment in the RE sector target? The unequivocal answer to this question is, of course, the private sector. More specifically, this study will focus only on a very specific part, or a type of activity conducted by the private sector – foreign direct investment (FDI). The importance of FDI in the context of renewable energies will be explained, as well as the list of barriers that are currently inhibiting the efforts of the private sector in this area.

DOI: 10.4324/9780429277207-2

Lastly, and most briefly, this chapter will ask the 'what' question: what is it that public policies should aim to achieve? This part will explain the importance of setting targets that are not only ambitious, but also comprehensive. More specifically, it will identify the necessity to achieve integration of both *means* and *outcomes*, a process that will be addressed in more detail in Chapter 2 of this study.

This chapter will present the reader with a preliminary scientific and economic review. The reader should note that this is a fast-changing area and yesterday's science may not necessarily be valid at the time of reading. The main points presented in this chapter, however, will remain, unfortunately, correct also in years to come. And these points – this chapter's main findings – are:

- First, energy poverty is a *substantive, and urgent* problem in the Global South.
- Second, without public policies, this problem will not be resolved fast enough for the planet, or for those living without access to energy.
- Third, one important objective that these public policies will have to target, is the increase of private FDI in the RE sector, in the Global South.
- Fourth, there are significant barriers that are currently inhibiting the flow of FDI in the RE sector towards the Global South. These barriers will require tailored responses, in the shape of laws and policies.

The following chapter elaborates on these issues.

II Why? SDG 7 and beyond

The first question that must be answered (albeit briefly) before delving into the waters of international law, is 'why?' or why is it useful and necessary to adopt states' policies for the promotion of RE-related investment in the Global South? There are, after all, those who will argue in favour of avoiding expensive public intervention, and allowing the invisible hand of the free markets to guide investment in this area. Perhaps surprisingly, these arguments are intensifying rather than diminishing over time. Addressing the need for public policies and public intervention is therefore necessary.

A Why should states intervene?

Why then, should states adopt policies for the promotion of investment in the RE sector? This question cannot be ignored, especially as states' RE policies are often expensive and even controversial due to their impact on competing environmental goals (e.g. wildlife conservation, or the aesthetic impact on nature[1]). These controversies are intensifying in the context of development assistance, where resources are scarce and the trade-offs could be significant.

This 'why' question is in fact being asked by several different groups, each for its own reasons and motives. First, there are those who deny the impact and importance of climate change, and consequently reject the notion that public resources should be invested in addressing this phenomenon. The author has no wish, nor the capability to interact with such claims. This debate is not being conducted on the scientific and academic spheres, but mostly within the political and public worlds. For the sake of this study, the author will rely on the scientific consensus, and will leave this part of the debate for the likes of Al Gore, David Attenborough, Nigel Farage and Donald Trump.

Second, there are those who genuinely care for the environment, but take issue with the trade-offs that are involved in the adoption of expensive RE-related public policies. To begin with, this group claim that the falling prices of RE technologies do not justify expensive public investment and intervention. Certain RE technologies, it is claimed, are already competitive in market terms and therefore attract record investment. Why then, should states invest scarce resources in the promotion of RE, when the free market is perfectly capable of delivering this public good on its own? Some members of this group also add that the result of such public interventions will be the allocation of precious public funds (e.g. Official Development Assistance (ODA)) to wealthy energy corporations, instead of for the genuine needs of developing countries.

It is true that the prices of RE technologies have dropped significantly in recent decades. As a result, renewables have become far cheaper and increasingly competitive. For example, in just six years (2010–2016) the global price for solar photovoltaic (PV) energy and onshore wind power fell by 69% and by 18%, respectively.[2] In certain suitable locations (e.g. Dubai, Mexico, Saudi Arabia for PV, and Denmark, Germany and the UK for wind power) competitive RE prices are already accessible, and further reductions are expected.[3] Moreover, it cannot be denied that these reductions in prices are already resulting in a significant investment in RE-related projects. For

1 Committee on US-China Cooperation on Electricity from Renewable Resources, *The Power of Renewables: Opportunities and Challenges for China and the United States* (The National Academies Press 2010) <https://www.nap.edu/read/12987/chapter/6#101> 103–104.

2 This data is referring to the global weighted average cost from large-scale solar PV plant. See IRENA 'Renewable power: Sharply falling generation costs' (IRENA 2017) <https://www.irena.org/-/media/Files/IRENA/Agency/Publication/2017/Nov/%20IRENA_Sharply_falling_costs_2017.pdf>.

3 IRENA 'Renewable power: Sharply falling generation costs' (n 2) 2.

example, at the time of writing, global investment in solar PV is at a record level.[4] These developments are without doubt encouraging and do imply a decreasing role for public policy and intervention.

The following section will address these questions and explain why governmental intervention is (still) necessary in order to increase investment in the RE sector.

i Technological advancements are not enough to stop climate change

The technological developments discussed above are certainly promising. On their own, however, they will not be enough for achieving the objectives of the Paris Agreement ('well below 2°C [...] pursuing efforts to limit the temperature increase to 1.5°C above pre-industrial levels'). The transition into a green economy is simply not happening fast enough[5]: according to the International Energy Agency (IEA), reaching the Paris Agreement's objective will 'require an energy transition of exceptional scope, depth and speed [... and] require an unparalleled ramp up of all low-carbon technologies in all countries'.[6] Concerning the urgency involved in the process of transitioning into RE-based economy, the IEA states[7]:

> Improvements to energy and material efficiency, and higher deployment of renewable energy are essential components of any global low-carbon transition. In the 66% 2°C Scenario, aggressive efficiency measures would be needed to lower the energy intensity of the global economy by 2.5% per year on average between 2014 and 2050 (three-and-a-half times greater than the rate of improvement seen over the past 15 years); wind and solar combined would become the largest source of electricity by 2030. This would need to be accompanied by a major effort to redesign

4 IEA, *World Energy Investment 2018* (IEA 2018) 48.

5 Dirk Röttgers 'The government role in mobilising investment and innovation in renewable energy' (2017) OECD Investment Insights. <https://www.oecd.org/investment/The-government-role-in-enabling-investment-and-innovation-in-renewable-energy.pdf>; Dolf Gielen et al. 'The role of renewable energy in the global energy transformation' (2019) 24 *Energy Strategy Reviews* 38, 39.

6 IRENA/IEA, *Perspectives for the Energy Transition: Investment Needs for a Low-carbon Energy System* (IEA/IRENA 2017) 7. For a review of the reasons for the expected lengthy timeframe, see Vaclav Smil 'Examining energy transitions: A dozen insights based on performance' (2016) 22 *Energy Research & Social Science* 194. For more optimistic views, see Florian Kern and Karoline Rogge 'The pace of governed energy transitions: Agency, international dynamics and the global Paris Agreement accelerating decarbonisation processes? (2016) 22 *Energy Research & Social Science* 13. Kern and Rogge's paper was written in 2016 and much of their optimism is influenced by the Paris Agreement which was just concluded at the time. Historic perspective, however, may suggest that their optimism (at least with respect to the Paris Agreement) was perhaps misplaced.

7 IRENA/IEA, *Perspectives for the Energy Transition* (n 6) 8.

electricity markets to integrate large shares of variable renewables, alongside rules and technologies to ensure flexibility.

Moreover, current states' policies (as appear in states' Nationally Determined Contributions (NDCs)) are insufficient.[8] An International Renewable Energy Agency (IRENA) report (as well as other sources[9]) states that even if implemented in full, current NDCs are still far from resulting in sufficient cuts in emission levels.[10] This report adds that '[u]nder today's national energy plans, renewable energy would bring little change in the supply mix over this time frame, since those plans mainly reflect market trends'.[11] Reaching IRENA's Renewable Energy Roadmap (REmap[12]) target of two-thirds RE share by 2050 will require a significant, and much faster increase in the share of RE – somewhere between six[13] and eight times[14] faster in comparison to the previous years' pace. Additional investment will also be necessary. Different sources provide different estimates regarding the exact additional sum that will have to be invested; IRENA estimates that an additional USD 7 trillion will have to be invested by 2050.[15]

While the need for increased investment is clear, a 2018 IEA report suggests that this is not necessarily happening.[16] This report points at 'a pause

8 Signe Krogstrup and William Oman 'Macroeconomic and financial policies for climate change mitigation: A review of the literature' (2019) IMF Working Paper WP/19/185 <https://www.imf.org/en/Publications/WP/Issues/2019/09/04/Macroeconomic-and-Financial-Policies-for-Climate-Change-Mitigation-A-Review-of-the-Literature-48612> 12.

9 The gap between states NDCs commitments and the necessary emission reductions is evaluated yearly by the UNEP in their annual Emissions Gap Report.

10 According to IRENA,

> It should be noted that the energy plans for 40% of the world's total primary energy supply were not consistent with the corresponding NDCs. If all energy plans matched the NDCs, CO_2 emissions in the Reference Case would be about 3 Gt per year lower in 2030. Fully implementing the NDCs, therefore, is a welcome first step. However, they alone are not enough. If implemented in full, they will reduce emissions by 20% by 2030, far less than the 50% decline that is needed.

IRENA/IEA, *Perspectives for the Energy Transition* (n 6) 129.

11 IRENA/IEA, *Perspectives for the Energy Transition* (n 6) 137.

12 'REmap explores the energy transition to decarbonise the energy system in line with the goal in the Paris Agreement of limiting global temperature rise to less than 2o C above pre-industrial levels with a 66% probability'. IRENA/IEA, *Perspectives for the Energy Transition* (n 6) 125.

13 IRENA *Global Energy Transformation: A Roadmap to 2050* (IRENA 2018) 25; Dolf Gielen et al. (n 5) 48.

14 IRENA/IEA *Perspectives for the Energy Transition* (n 6) 137.

15 IRENA, *Roadmap to 2050* (n 13) 11. Dasgupta et al. mention other estimates, including USD 6.78 trillion a year in order to meet the 2°C scenario. Dipak Dasgupta, Jean-Charles Hourade and Seyni Nafo, *A Climate Finance Initiative: To Achieve the Paris Agreement and Strengthen Sustainable Development* (2019) <http://www2.centre-cired.fr/IMG/pdf/rapport_gicf.pdf> 3.

16 IEA, *World Energy Investment 2018* (IEA 2018) 12.

in the shift of investments towards cleaner sources of energy supply', as global investment in renewables declined in 2017 by 7%. One of the IEA report's authors has referred to this decline as a 'warning signal for clean energy transition'.[17] An exception to this decline is solar energy, which reached a new record investment in 2017. This increase in investment was attributed mostly to state-directed investment in China, which, according to the IEA, confirms the important role of public intervention and policies.[18]

Indeed, where public support has weakened over the years, investment has fallen as well. The UK is an interesting case study in this respect. Extremely successful public support schemes have led to impressive achievements. A third of the UK's electricity is generated from renewable sources[19] and the country is well underway to meet its next carbon budget. This success, however, is attributed to past policies; many of which are no longer in place.[20] The House of Commons Environmental Audit Committee pointed in a report to a significant reduction in investment, stating that 'Annual clean energy investment in the UK is now the lowest it has been since 2008'.[21]

Lastly, it is important to remember that even if prices of RE are reducing and the share of RE investment is on the rise, investment in fossil fuels is still happening, and on a very large scale. Every new investment in fossil fuel facilities de facto ensures that states will continue to emit greenhouse gases (GHG) for decades, as far as 40 years to come.[22] The free market's invisible hand may indeed lead, eventually, to a full transition into a greener economy. But until then, many more fossil fuel facilities will be built and decades of pollution will be guaranteed. The urgency of acting *now*, and ensuring that states will not be locked into a polluting future,[23] is therefore clear.

17 Michael Waldron 'Commentary: Decline in renewable investment is a warning signal for clean energy transitions' (24 July 2018) *IEA News* <https://www.iea.org/newsroom/news/2018/july/commentary-despite-falling-costs-the-decline-in-renewables-investment-is-a-warn.html>.

18 IEA 'Global energy investment in 2017 fails to keep up with energy security and sustainability goals' (17 July 2018) *IEA News* <https://www.iea.org/newsroom/news/2018/july/global-energy-investment-in-2017-.html>.

19 See data published by the UK government in April 2019 'Energy trends: Renewables' <https://assets.publishing.service.gov.uk/government/uploads/system/uploads/attachment_data/file/789370/Renewables_March_2019.pdf>.

20 House of Commons Environmental Audit Committee *Green Finance: Mobilising Investment in Clean Energy and Sustainable Development*. Sixth Report of Session 2017–2019 (16 May 2018) <https://publications.parliament.uk/pa/cm201719/cmselect/cmenvaud/617/617.pdf> 10.

21 House of Commons Environmental Audit Committee (2018) (n 20), 9.

22 Jan Christoph Steckel and Michael Jacob 'The role of financing cost and de-risking strategies for clean energy investment' (2018) 155 *International Economics* 19.

23 Ottmar Edenhofer et al. 'Closing the emission price gap' (2015) 31 *Global Environmental Change* 132.

ii Government intervention increases technological innovation

The role that technological innovation is expected to play in the abatement of climate change is crucial.[24] In order to limit the concentration of atmospheric CO_2 to 500 parts per million (PPM) by 2095, it has been argued that fossil fuel-based technologies must be completely removed by that year.[25] More research has shown that the attainment of the 1.5°C goal is possible if current carbon-intensive infrastructure will be phased out at the end of its designed lifetime.[26] In this respect, the gap between what should be done, and what is currently realistically feasible, is expected to be bridged by new efficient technologies. It is clear therefore that 'any sensible climate policy will have technological development as a main focus'.[27] Indeed, the 2018 Intergovernmental Panel on Climate Change (IPCC) report included technological innovation as one of the necessary factors for preventing a rise of more than 1.5°C[28]; a fact that was also recognised in international agreements.[29]

Public support and intervention is crucial for the development of new clean technologies. Indeed, research has shown a clear correlation between climate support schemes (including 'technology push' policies such as public investment in R&D, and 'market pull' policies such as Feed-in Tariffs (FIT)), and the registration of patents for clean technologies.[30]

iii Government intervention is necessary for reflecting the real cost of investments

Public policies and intervention are also necessary for reflecting the cost of energy sources more accurately. The issue of attributing the correct price to

24 UNFCCC, Subsidiary Body for Scientific and Technological Advice, *Report on Options to Facilitate Collaborative Technology Research and Development*, UN DOC. FCCC/SBSTA/2010/INF.11 (2010), online: UNFCCC <http://unfccc.int/resource/docs/2010/sbsta/eng/inf11.pdf> [SBSTA 2010]; Knut H. Alfsen and Gunnar S. Eskeland 'A broader palette: The role of technology in climate policy' (2007) *Report to the Expert Group for Environment Studies* 2007, 1 <http://www.regeringen.se/content/1/c6/07/89/07/788128bb.pdf> 38–40.

25 Leon Clarke, Katherine Calvin, Jae Edmonds, Page Kyle and Marshall Wise 'When technology and climate policy meet: Energy technology in an international policy context' (2008) *Harvard Project on International Climate Agreements, Discussion Paper* 2008–2021 <http://belfercenter.ksg.harvard.edu/files/ClarkeFinalRevised.pdf> 16.

26 Christopher Smith et al. 'Current fossil fuel infrastructure does not yet commit us to 1.5°C warming' (2018) *Nature Communications* <https://www.nature.com/articles/s41467-018-07999-w>.

27 Alfsen and Eskeland, (n 80) 45.

28 IPCC, *Global Warming of 1.5°C: Executive Summary for Policy Makers* (2018) 23.

29 See, for example, Art 10 of the Paris Agreement.

30 Geraldine Ang et al. 'The empirics of enabling investment and innovation in renewable energy' (2017) OECD Environment Working Paper No. 123, 11, 79.

any given source of energy is well known and often discussed.[31] In essence, under current markets' terms, energy sources are not priced correctly. The *real* price of renewables should also reflect very significant economic benefits, notably the prevention of expected damage that is caused by climate change. Equally, the *real* price of fossil fuels should include the expected damage that such energy sources are creating, not to mention other distorting factors such as fossil fuel subsidies.[32] A *genuine* risk/balance calculation will improve the likelihood of investment in places that are currently deemed as too risky in light of expected revenues. This could happen only through government interventions and policies. If added to investors' decisions risk/benefit calculation, these unaccounted-for benefits could change, and even tip, the balance in favour of RE.

Moreover, the need to intervene in order to correct the business-case for renewables is especially striking when one considers the resources that are available in the developing world. For example, Africa is abundant with renewable resources. As reviewed by Ouedraogo, Africa's *potential* generation capacity from hydropower is about 1,200 Terawatt hours (TWh).[33] The *potential* generation capacity from Concentrated Solar Power (CSP) and PV is 470,000 TWs, and 660,000 TWh, respectively.[34] From wind power, Africa could *potentially* generate 460,000 TWh.[35] Despite these substantial resources, only a small fraction of Africa's energy consumption arrives from RE, and the potential for further development is significant.

iv Investment in the developing world: additional barriers, and gains

Lastly, and perhaps most importantly, there are also the issues of location and development. There is no doubt that significant advancements in RE technologies are indeed being translated into substantial investment in renewables. The markets are certainly delivering on this front (even if not quickly enough to match the urgency that is embedded in climate change[36]). It is nevertheless important to note that most investments are taking place

31 Ottmar Edenhofer et al. (n 23); Signe Krogstrup and William Oman (n 8) 14–15.

32 According to the IEA fossil fuel consumption subsidies are increasing, totalling more than USD 300 billion per year in 2017. A REN21 report estimated this number at USD 370 billion. See IEA <https://www.iea.org/weo/energysubsidies/> and REN21 <http://www.ren21.net/gsr-2018/pages/highlights/highlights/#page-content>.

33 Nadia Ouedraogo 'Opportunities, barriers and issues with renewable energy development in Africa: A comprehensible review' (2019) *Current Sustainable/Renewable Energy Reports* 1, 2.

34 Ouedraogo (n 33) 2; Ehsanul Kabir et al. 'Solar energy: Potential and future prospect' (2018) 82 *Renewable and Sustainable Energy Reviews* 894, 895.

35 Ouedraogo (n 33) 2.

36 Gregor Semieniuk and Mariana Mazzucato 'Financing green growth' (2018) Working Paper IIPP WP 2018-04 <https://www.ucl.ac.uk/bartlett/public-purpose/sites/public-purpose/files/iipp-wp-2018-04.pdf>.

in very specific locations, notably in China and the developed world. This study, however, evaluates developments in the Global South where, while investment needs are great,[37] actual investment is scarce, and energy poverty is a significant problem.[38]

In the developing world, as reviewed in this study, the barriers for investment are much wider and far more complex than the mere price of technologies. Other factors, such as the high cost of finance and political risks, are playing a crucial role and inhibiting investment.

It is important to stress in this respect that the urgency to increase RE-related investment in the Global South should be understood beyond the context of climate change. First, there is the significant energy poverty that still exists. More than 1 billion persons – 14% of the world's population – do not have access to energy.[39] In most African states the overall rate of access to energy is below 50%,[40] and only 38.8% of the population in low-income countries have access to electricity.[41] In rural areas the numbers are even lower: only 28% of low-income states' rural populations (24.8% in the sub-Saharan region) have access to electricity.[42] Only about 60% of the world's population are relying on clean fuels for cooking,[43] and in African rural areas 92% of the population rely on fuelwood, charcoal, dung and agricultural residues for this purpose.[44]

Importantly, the context of Global South countries requires governmental intervention that will be much wider in scope.[45] As stated in the literature, targeted interventions in the form of carbon pricing or support schemes were successful mostly in the developed world and emerging economies.[46] Developing markets suffer from a wide range of barriers and will require far more ambitious intervention, whether through international laws (as reviewed in this study) or domestic reforms in areas ranging from land law to public procurement, energy market regulations to competition laws.

SDG 7 indeed instructs the international community to significantly improve access to clean energy, especially in the developing world.[47] Increasing

37 According to Dasgupta et al. 63% of low-carbon investment will have to take place in the developing world, Dipak Dasgupta, Jean-Charles Hourade and Seyni Nafo (n 15) 3.

38 For example, according to the IEA sub-Saharan states account for 14% of the world's population, but manage to attract only 4% of the world's investment in energy. IEA, *Energy Access Outlook 2017: From Poverty to Prosperity* (IEA 2017) 76.

39 IEA, *Energy Access Outlook 2017* (n 38) 49.

40 IEA, *Energy Access Outlook 2017* (n 38) 80–81.

41 See World Bank's data <https://data.worldbank.org/indicator/eg.elc.accs.zs>.

42 See World Bank's data <https://data.worldbank.org/indicator/EG.ELC.ACCS.RU.ZS>.

43 See World Bank's data <https://data.worldbank.org/indicator/eg.cft.accs.zs>.

44 IEA, *Energy Access Outlook 2017* (n 38) 91.

45 Dipak Dasgupta, Jean-Charles Hourade and Seyni Nafo (n 15).

46 Röttgers (n 5) 2–3.

47 UNGA, *Transforming Our World: The 2030 Agenda for Sustainable Development* (Resolution adopted on 21 October 2015).

RE-related investment (e.g. through off-grid and standalone grid systems) is a straightforward answer to this call. Moreover, addressing energy needs will also result in important ancillary impacts. As stated by others,[48] SDG 7 is not only an objective, but also an important tool that is expected to enhance the attainment of almost all other SDGs. Increased access to clean energy will address a myriad of issues, ranging from hunger (SDG 2), health (SDG 3), economic growth (SDG 8), industry and infrastructure (SDG 9), climate action (SDG 13) and more.[49] The trade-off that is embedded in investing resources in RE-related investment should, therefore, take into consideration a wide variety of gains.

These barriers are creating a second layer for the challenge discussed in this study, which seems like a direct conflict between SDG 7 (access to energy) and SDG 13 (climate action). It could be that a more efficient, immediate reduction in emissions will be achieved if resources will be directed towards the developed world, or at least to middle-income states, where investment conditions and investment risks are preferable over conditions in the developing world. Indeed, as reviewed in Chapter 6 of this study, authors have identified that states' climate-related Official Development Assistance is directed mostly to middle-income states and only rarely to Least Developed Countries, where energy poverty is far more endemic.[50] Continuing with business as usual will not lead to progress with respect to the attainment of SDG 7.

In summary, it is clear that public policies and intervention are necessary for increasing investment in the RE sector. Without these policies, challenges such as climate change and energy poverty will not be met – at least not fast enough for the planet and those most deprived among us.

III Who? Private foreign direct investment in the RE sector

The next question that should be discussed is 'who?' Or, more specifically, who should be targeted by public policies? The clear and unequivocal

48 Gregor Schwerhoff and Mouhamadou Sy 'Financing renewable energy in Africa – key challenge of the sustainable development goals' (2017) 75 *Renewable and Sustainable Energy Reviews* 393; Nadia Ouedraogo 'Opportunities, barriers and issues with renewable energy development in Africa: A comprehensible review' (2019) 2 *Current Sustainable/Renewable Energy Reports* 1.

49 See, for example, impact on jobs at Nicola Cantore et al. 'Promoting renewable energy and energy efficiency in Africa: A framework to evaluate employment generation and cost effectiveness' (2017) 12 *Environmental Research Letters*.

50 Homi Kharas 'Aid and climate finance' (2015) *Brookings* <https://www.brookings.edu/wp-content/uploads/2016/07/aid-climate-finance-kharas-1.pdf> 26; Giorgio Gualberti and Lorcan Lyons, *Missing the mark: Gaps and lags in disbursement of development finance for energy access* (Sustainable energy for all 2017) <https://www.afdb.org/fileadmin/uploads/afdb/Documents/Publications/Missing_the_Mark_Gaps_and_Lags_in_Disbursement_of_Development_Finance_for_Energy_Access.pdf> 8.

answer to this question is the private sector. Put simply, the involvement of the private sector is crucial for climate change mitigation and adaptation. In no area is this more evident than in the case of investment in renewable energy. Below is a short review of the role and importance of the private sector's investment in RE.

A The role of the private sector

The need to scale up the role of the private sector in supporting the shift into a greener, RE-based economy was recognised by the international community on numerous occasions. For example, the Bali Action Plan[51] provides that the Parties are encouraged to consider ways to strengthen the involvement of the private sector in the mitigation of climate change.[52] The Copenhagen Accord emphasises the use of markets and incentives in this context,[53] as well as the fact that the funding pledged to the developing world (USD 100 billion per year) is expected to arrive *inter alia* from private sources. The Cancun Agreements continued this line of action as decision CP.16 (titled: 'Financial mechanisms of the Convention') requested the Global Environmental Facility ('GEF') to continue *inter alia* 'promoting private-sector financing and investment to address climate change activity'.[54] Similarly, the Paris Agreement stresses the importance of incentivising and facilitating the participation of 'private entities' in the mitigation of greenhouse emissions.[55] The UNFCCC Decision 1/CP.21 ('Adoption of the Paris Agreement') further emphasises the role of 'Non-Party Stakeholders',[56] and 'welcomes the efforts of all non-Party stakeholders [...]' including the private sector, and calls on these to 'scale up their efforts'.[57]

And what role is the private sector expected to fulfil? The private sector's contribution to the global effort is being done in numerous ways. Notable types of 'private' activities which support climate mitigation and adaptation efforts include international trade, investment (portfolio and FDI), finance,

51 UNFCCC, *Bali Action Plan*, COP Decision 1/CP.13, UN Doc. FCC/CP/2007/6/Add.1 (2008).

52 The Bali Action Plan, *supra* note 51 at Articles 1(b)(vii), (c)(v), (e)(v). See also later international accords and decisions as described above, for example, Article 7–8 of the *Copenhagen Accord, infra* note 53.

53 Copenhagen Accord, UNFCCC, *Copenhagen Accord,* COP Dec. 2/CP.15, UN FCCCOR, UN Doc. FCCC/CP/2009/11/Add.1 (2009).

54 Article 4(f) of UNFCCC, *Financial mechanism of the Convention: Fourth, review of the financial mechanism,* CP.16 (2010), online: UNFCCC <https://unfccc.int/files/meetings/cop_16/conference_documents/application/pdf/20101204_cop16_fm.pdf>.

55 UNFCCC, *Paris Agreement,* UN Doc. FCCC/CP/2015/10/Add.1, Annex, [*Paris Agreement*], in Article 6(4)(b).

56 UNFCCC, *Adoption of the Paris Agreement, Decision 1/CP.21,* UN Doc. FCCC/CP/2015/10/Add.1, in Part V.

57 *Adoption of the Paris Agreement* (n 56) paras 133–134.

insurance, research and development and more.[58] This chapter's focus, however, is primarily on the private FDI in the RE sector.

Private FDI is important for mitigation and adaptation policies for several reasons.[59] Two notable, often mentioned areas in which FDI are crucial in this respect are (1) climate finance, and (2) the development of climate-related technologies, and their assimilation in the developing world (via the process of technology transfer). While a detailed review of the manner in which FDI is supportive of climate finance and technologies is beyond the scope of this chapter, a brief overview is offered in order to provide the necessary background, based on which the second part of this study will be founded.

B FDI and climate finance

The role that FDI is expected to play in the efforts to achieve the SDGs is widely acknowledged.[60] More specifically with respect to the topic of this study, Chapter 16 of the 2014 IPCC report (investment and finance) opens with the following caveat[61]:

> Scientific literature on investment and finance to address climate change is still very limited and knowledge gaps are substantial; there are no agreed definitions for climate investment and climate finance.

58 See among other sources UNFCCC, *Trends in Private Sector Climate Finance: Report Prepared by the Climate Change Support Team of the United Nations Secretary General on the Progress Made since the 2014 Summit* (UNFCCC 2015) <http://www.un.org/climatechange/wp-content/uploads/2015/10/SG-TRENDS-PRIVATE-SECTOR-CLIMATE-FINANCE-AW-HI-RES-WEB1.pdf>; Thomas Cottier, Olga Nartova and Sadeq Bigdeli (eds.) *International Trade Regulation and the Mitigation of Climate Change* (Cambridge University Press 2009); WTO/UNEP, Ludivine Tamiotti et al. *Trade and Climate Change: WTO-UNEP Report* (WTO 2009); UNCTAD, *World Investment Report 2010, Investing in a Low Carbon Economy* (United Nations Publications 2010).

59 See for example reasons stated in OECD, *Overcoming Barriers to International Investment in Clean Energy* (OECD 2015) <http://www.oecd-ilibrary.org/environment/green-finance-and-investment_24090344> 16.

60 See for example in the Addis Ababa Action Agenda of the Third International Conference on Financing for Development (UN 2015), para 44:

> We recognize the important contribution that direct investment, including foreign direct investment, can make to sustainable development, particularly when projects are aligned with national and regional sustainable development strategies. Government policies can strengthen positive spillovers from foreign direct investment, such as know-how and technology, including through establishing linkages with domestic suppliers, as well as encouraging the integration of local enterprises, in particular micro, small and medium-sized enterprises in developing countries, into regional and global value chains.

61 Gupta et al. 'Cross-cutting investment and finance issues. In: Climate change 2014: Mitigation of climate change' in *Contribution of Working Group III to the Fifth Assessment Report of the Intergovernmental Panel on Climate Change* (IPCC 2014) Chapter 16, 1210.

> Quantitative data are limited, relate to different concepts, and are incomplete. Accounting systems are highly imperfect.

Despite the obvious uncertainties, there is no doubt that increasing private FDI is crucial for meeting the challenge of climate finance.[62] In a nutshell, despite an encouraging increase in investment,[63] current investment levels are simply too low and the transition into a green economy is not happening fast enough. The International Energy Agency commented on this situation[64]:

> From an investment perspective, the energy sector transition in the 450 scenario requires not only more capital investment in energy, but also a different allocation of capital. National policies have not been sufficient, in the absence of a global climate agreement, for this shift in investment flows to take place at the necessary rate and speed.

In 2018 BloombergNEF estimated that a total of USD 332 billion was invested in low-carbon technologies.[65] According to the IEA, in order to achieve the 450 PPM scenario investment in low-carbon technologies, this must increase to USD 400 billion by 2030.[66] The IEA further estimate that in order to meet the 450 PPM goal, 80% of the overall investment in energy should be in low-carbon technologies after 2020, and 90% after 2025.[67] Indeed the hunger for finance in clean energy is striking; as estimated by the IEA, meeting the pledges submitted by the Member States of the UNFCCC in COP 21 alone will require a USD 13.5 trillion investment in energy efficiency and low-carbon technologies from 2015 to 2030.[68]

As stated above, in several international agreements it was indeed mentioned that at least a part of the promised financial assistance that was

62 IPCC, *Global Warming of 1.5°C: Summary for Policy Makers* (2018), 23 D.5.1.

63 According to the Frankfurt School-UNEP report, USD 2.2 trillion were invested in renewables in the years 2010–2017. Frankfurt School-UNEP, Global Trends in Renewable Energy Investment 2018.

64 International Energy Agency, *Special Report: World Energy Investment Outlook* (IEA 2014) 40.

65 BloombergNEF, *Clean Energy Investment Trends 2018* (Bloomberg 2018) <https://data.bloomberglp.com/professional/sites/24/BNEF-Clean-Energy-Investment-Trends-2018.pdf>.

66 International Energy Agency, *Energy and Climate Change* (OECD/IEA 2015) 67.

67 IEA 'The way forward: Five key actions to achieve a low-carbon energy sector' see discussion of action 3 <http://www.iea.org/publications/freepublications/publication/The_Way_forward.pdf>. IEA data indicates a share of only 66% in 2017, see IEA, *World Energy Investment 2018* (n 4) 14.

68 IEA, *Energy and Climate Change: World Energy Outlook Special Briefing for COP 21* (IEA 2015) <http://www.worldenergyoutlook.org/media/news/WEO2015_COP21 Briefing.pdf> 4.

pledged by the developed world should arrive from private sources.[69] Also the UN Environment Program (UNEP) has estimated that the vast majority (about 80%) of the investment needed to fund the global transition into the 'green economy' will arrive from the private sector.[70] It is clear therefore, that there is a great need for increasing climate finance, and that much of it is expected to arrive from the private sector.[71]

There are many sources of private finance, including international investment (portfolio and FDI), payment for carbon credits,[72] as well as money raised through capital markets.[73] Among these sources, FDI (mostly project developers[74]) is considered as one of the most important tools for the mobilisation of climate finance, being the largest source of North-South climate-related financial flow.[75] The relevance, and the importance, of FDI for climate finance, is therefore clear.

As stated above, the need for investment in RE in the developing world goes beyond addressing climate change. Such investment will also improve acute problems such as access to energy, and indirectly support a range of other issues, such as health, education, gender equality and more. Seen through the lens of development, it seems clear that the aim of policies should not be the mere scaling up of investment, but mostly the scaling up of investment in the Global South. According to data, up to 92% of private investments are made locally, within the same countries in which the investors reside (mostly China, Europe, Japan and the US).[76] Only a small share of investment in infrastructure within developing countries currently arrives from the private sector,[77] and only a very

69 See review above, and more specifically the discussed Copenhagen Accord and the Cancun Agreements.

70 UNEP, *Green Economy Report: Finance, Supporting the Transition to a Global Green Economy* (UNEP 2011) 586. This is hardly surprising, as private sources currently represent about 62% of global climate finance, see in Barbara Buchner et al., *Global Landscape of Climate Finance 2015* (CPI 2015) <http://climatepolicyinitiative.org/wp-content/uploads/2015/11/Global-Landscape-of-Climate-Finance-2015.pdf> 1.

71 The private sector is already the main investor in RE. According to IRENA, 90% of investment are arriving from the private sector. IRENA, *Global Landscape of Renewable Energy Finance 2018* (IRENA 2018).

72 See in Martin Stadelmann, Axel Michaelowa and J.Timmons Roberts 'Difficulties in accounting for private finance in international climate policy' (2013) 13(6) *Climate Policy* 718, 721–728.

73 Christa Clapp, Jane Ellis, Julia Benn and Jan Carfee-Morlot, *Tracking Climate Finance: What and How?* (OECD 2012) <http://www.oecd.org/env/cc/50293494.pdf> 13.

74 Buchner et al. 2015 (n 70) 5.

75 Martin Stadelmann and Axel Michaelowa, *Contribution of the Private Sector to Climate Change Long-Term-Finance: An Assessment of Private Climate Finance Mobilized by Switzerland* (FOEN 2013) 10.

76 Buchner et al. (n 70) 10; IRENA, *Global Landscape of Renewable Energy Finance 2018* (n 71).

77 USD 146 billion, out of USD 1 trillion overall investment. Oliver Johnson et al. 'Catalysing investment in sustainable energy infrastructure in Africa: Overcoming financial and non-financial constraints' (2017) SEI Working Paper No 2017-03 <https://mediamanager.

small portion of this share (7%) is invested in the poorest nations within the developing world.[78]

The above suggests that the vast majority of the developing world is still not benefiting from this much-needed finance. The potential available in FDI, as a tool for mobilising the much-needed climate finance to the developing world, remains for the time being, locked.

C FDI and climate technologies global assimilation

A second element in which the importance of FDI is demonstrated is the need to ensure the transfer of advanced green technologies from a relatively small group of developed states in which these technologies are being developed, to the developing world.

The process of technology transfer was described by the UNFCCC as 'a broad set of processes covering the flows of know-how, experience and equipment for mitigating and adapting to climate change among different stakeholders'.[79] The role of the development and transfer of technology in the abatement of climate change is crucial.[80] This role is emphasised by the fact that the global energy usage is expected to (at least) double itself by the end of the century,[81] and a mere change in life habits (i.e. action such as better house insulating, increased usage of public transportation) can reduce emissions by only 10–20% (under optimistic scenarios).[82]

The need for technology transfer was recognised by the international community *inter alia* under Article 4 of the UNFCCC ('Commitments'), in which it was stated that the Parties shall:

---

sei.org/documents/Publications/SEI-WP-2017-03-Africa-energy-infrastructure-finance.pdf> 5.

78 This data refers to the first half of this data and is likely to be somewhat dated by the time of publication. The overall 'trend' of very low investment, however, remains very much the same. Johnson et al. (n 77) 6.

79 UNFCCC Glossary <http://unfccc.int/essential_background/glossary/items/3666.php#T>.

80 UNFCCC, Subsidiary Body for Scientific and Technological Advice, *Report on Options to Facilitate Collaborative Technology Research and Development*, UN DOC. FCCC/SBSTA/2010/INF.11 (2010) <http://unfccc.int/resource/docs/2010/sbsta/eng/inf11.pdf>; Knut H. Alfsen and Gunnar S. Eskeland 'A broader palette: The role of technology in climate policy' (2007) 2007:1 Report to the Expert Group for Environment Studies <http://www.regeringen.se/content/1/c6/07/89/07/788128bb.pdf> 38–40.

81 Alfsen and Eskeland, ibid. 41, elsewhere the authors present the option that energy consumption will treble itself by 2050. See Alfsen and Eskeland (n 80) 43–44.

82 Ibid. at 38. The need to invest in new technologies was stressed in the 2014 IPCC report stating that 'there is wide agreement among model results on the necessity to ramp up investments in research and development (R&D)' [...], estimating additional needed funding between USD 4.5 and 78 billion per year during 2010–2029, and USD 115–129 billion per year between 2030 and 2049. See IPCC, Chapter 16 (n 61) 1220.

> Promote and cooperate in the development, application and diffusion, including transfer, of technologies, practices and processes that control, reduce or prevent anthropogenic emissions of greenhouse gases [...][83]

Article 4(5) of the UNFCCC further states:

> The developed country Parties and other developed Parties included in Annex II shall take all practicable steps to promote, facilitate and finance, as appropriate, the transfer of, or access to, environmentally sound technologies and know-how to other Parties, particularly developing country Parties, to enable them to implement the provisions of the Convention. In this process, the developed country Parties shall support the development and enhancement of endogenous capacities and technologies of developing country Parties. Other Parties and organisations in a position to do so may also assist in facilitating the transfer of such technologies.[84]

The most important factor in the development of climate-friendly technology is, unsurprisingly, the country's general capacity for innovation.[85] Indeed, past research suggests that the bigger part of climate-friendly technology is developed within Organisation for Economic Co-operation and Development (OECD) countries,[86] and especially in Japan, South Korea, Germany, the US, and more recently also in China.[87] Innovation in other parts of the world is rather limited, and in places such as Africa can be described as almost non-existent.[88]

These findings emphasise the fact that technology is being developed only in certain parts of the world. Considering the above reviewed importance of technology, and the necessity of a common effort by *all* members of the international community, one inevitable conclusion arises: green technology must be globally transferred. Unfortunately, however, climate-friendly technologies do not seem to reach global distribution. Climate-friendly technologies

83 Article 4(1)(c) of the UNFCCC. See also Article 10 of the 2015 UNFCCC Paris Agreement.

84 Article 4(5) of the UNFCCC.

85 Ivan Hascic, Nick Johnstone, Fleur Watson and Chris Kaminker 'Climate policy and technological innovation and transfer: An overview of trends and recent empirical results' (2010) OECD Environment Working Papers No.30 <http://www.oecd-ilibrary.org/docserver/download/fulltext/5km33bnggcd0.pdf?expires=1297805911&id=0000&accname=guest&checksum=52ED5E1A3D682A6C30061B8C86CDD8DA> 44.

86 Ibid., 44.

87 Kristina Lybecker and Sebastian Lohse, *Innovation and Diffusion of Green Technologies: The Role of Intellectual Property and Other Enabling Factors* (WIPO 2015) 7; Antoine Dechezlepretre, Matthieu Glachant, Ivan Hascic, Nick Johnstone and Yann Meniere 'Invention and transfer of climate change technologies on a global scale: A study drawing on patent data' (2009) <http://papers.ssrn.com/sol3/papers.cfm?abstract_id=1414227>.

88 WIPO 2015 (n 87) at box 1 (p. 8).

are being transferred mainly within the developed world: 73% of these inventions' exportation was aimed at other developed countries, while only 22% of these travelled to emerging economies.[89] This data is supported by the fact that technology owners, who usually bear the costs of technology transfer, are incentivised to transfer their technologies only for profitable markets.[90]

FDI is often mentioned as an efficient, important[91] and proven[92] tool for technology transfer; indeed the OECD's inter-governmental Freedom of Investment Roundtable forum described FDI as a 'vital source of finance and a powerful vector of innovation and technology transfer…'.[93] However, as stated above, up to 92% of private direct investments are made locally,[94] effectively locking novel technologies within those developed states in which they were developed. Also, from the little private investment that does flow internationally, only a small fraction arrives in the Least Developed Countries.[95] The potential role that FDI can play in this respect is not being fulfilled.

D Barriers for private investment in RE

The barriers standing before RE-related investment are diverse and vary according to the type of investment, activity and location. Many institutions have attempted to map these barriers. A World Business Council for

89 Dechezlepretre et al. (n 87) 35. This data is admittedly dated, and the rise of China as a hub for climate innovation is relevant. The main picture remains unchanged – i.e. technology does not travel enough to the developing parts of the world.

90 UNFCCC, *Recommendations on Future Financing Options for Enhancing the Development, Deployment, Diffusion and Transfer of Technologies Under the Convention: Report by the Chair of the Expert Group on Technology Transfer* (2009) UN Doc. FCCC/SB/2009/2, para 52.

91 Tim Forsyth, *International Investment and Climate Change: Energy Technologies for Developing Countries* (3rd ed., Earthscan 2013); Kamal Saggi 'Trade, Foreign Direct Investment and International Technology Transfer: A Survey' (2002) 17(2) The World Bank Research Observer <http://www2.econ.uu.nl/users/marrewijk/pdf/ihs/student%20present/kamalsaggi.pdf> 207; Wolfgang Keller 'International Technology Diffusion' (2004) 42(3) *Journal of Economic Literature* 752, 769; Wolfgang Keller 'International Trade, Foreign Direct Investment, and Technology Spillovers' (2009) Working Paper 15442, NBER Working Paper Series, NBER <http://www.nber.org/tmp/88248-w15442.pdf>; Matthieu Glachant et al., *Promoting the International Transfer of Low-Carbon Technologies*, Report for the Commissariat general a la strategy et a la prospective (2013) <http://personal.lse.ac.uk/dechezle/Promoting_the_international_transfer_of_low_carbon_techs.pdf>.

92 Theodore Moran 'How to investigate the impact of foreign direct investment on development and use the result to guide policy' (2007) *Brooking Trade Forum*, <http://siteresources.worldbank.org/EXTEXPCOMNET/Resources/2463593-1213989126859/12_EC_BBL4_Moran_Feb_14_Brookings_Paper_on_FDI.pdf>.

93 OECD Freedom of Investment Roundtable, *Harnessing Freedom of Investment for Green Growth, April 2011* (OECD 2011) <http://www.oecd.org/dataoecd/12/32/47721398.pdf>.

94 Buchner et al. (n 70) 10. Barbara Buchner *et al. The Global Landscape of Climate Finance 2014* (CPI 2014), available online: <http://ecreee.wikischolars.columbia.edu/file/view/Buchner+2014+-+The+Landscape+of+Climate+Finance.pdf> 18.

95 Glachant et al. (n 91) 21.

Sustainable Development ('WBCSD') report states that with respect to low-carbon technologies, variables such as strong signals from governments (targets or regulation), institutional framework (stable policies, transparency, etc.), appropriate absorptive capacity, proper financial incentives and productive business engagement with governments, are all relevant and important for the private sector.[96] A UNEP report concerning renewable energies identifies as barriers for investment in renewable energy: the risky nature of such an investment (including regulatory and political risks); low financial incentives; high costs; market failures regarding investment in R&D; current regulation and technical structure of the energy markets and infrastructure, and certain sustainability criteria (biofuels, environmental effects of large hydropower facilities, etc.).[97] Other studies have identified more specific risks, with respect to specific technologies.[98]

The IPCC identified that '[a] main barrier to the deployment of low-carbon technologies is a low risk-adjusted rate of return on investment vis-à-vis high-carbon alternatives often resulting in higher cost of capital'.[99] An OECD working paper identifies several barriers with respect to investment in infra-structure for low-carbon investment, including inadequate returns and 'unmanageable risks'.[100] Under 'unmanageable risks', the authors mention policy and regulatory risks (e.g. lack of policy certainties), political risks (e.g. 'short-termism of politicians'), currency risks, technological risks (e.g. technology may become out-dated relatively fast) and more.[101] The IEA addressed more generally the obstacles faced by investors in the energy sector (rather than the RE sector). These barriers, however, are valid also for RE investors, and the IEA research indeed included these within the wider energy sector. The IEA mentions *inter alia*, political risks (e.g. the resilience and stability of the political and legal systems, the possibility of expropriation, the credibility of energy policies), economic risks (e.g. high price volatility, unstable currency rates and economic environment) and variety of

96 World Business Council for Sustainable Development, *Enabling Framework for Technology Diffusion: A Business Perspective* (WBCSD 2010) WBCSD <http://www.wbcsd.org/web/projects/energy/EF_WBCSD_final_low2.pdf> 2–3.

97 UNEP Green Economy Report: Renewable Energy (n 145) 226–233.

98 See for example Nadine Gatzert and Thomas Kosub 'Risks and risk management of renewable energy projects: The case of onshore and offshore wind parks' (2016) 60 *Renewable and Sustainable Energy Reviews* 982; Jing Hu et al. (n 108); World Bank, *A Sure Path to Sustainable Solar: Solar Deployment Guidelines* (2019) <http://pubdocs.worldbank.org/en/155991570472678574/A-Sure-Path-to-Sustainable-Solar-Guidelines.pdf> 9.

99 IPCC, *Climate Change 2014: Mitigation of Climate Change. Contribution of Working Group III to the Fifth Assessment Report of the Intergovernmental Panel on Climate Change* (CUP 2014) 1211.

100 Jan Corfee-Morlot et al. 'Towards a green investment policy framework: The case of low-carbon, climate resilient infrastructure' (2012) OECD Environmental Working Papers, No 48 <http://dx.doi.org/10.1787/5k8zth7s6s6d-en> 26.

101 Corfee-Morlot et al. (n 100) 26.

project-specific risks.[102] Elsewhere, the IEA recommended specifically with respect to RE that in order to increase investment, states must improve elements such as the predictability and reliability of long-term policies, and adopt regulatory frameworks that support cost-effective remuneration.[103]

Also, the Multilateral Investment Guarantee Agency (MIGA) has flagged in the past the risks involved in energy investment as negatively unique. MIGA's 2013 World Investment and Political Risk Report states: 'From a sectoral viewpoint, energy sector deals (such as power purchase agreements and independent power production) prove to be more at risk of contract failure than projects outside the sector [...]'.[104] Relying on pre-claims events[105] involving a breach of contract, MIGA's report sees Latin America and sub-Saharan Africa as especially risky destinations for energy-related investment ('specifically power').[106]

In order to simplify and explain this rather hectic list of elements, the IPCC has grouped these factors into two main types (or groups) of barriers: *economic risks* and *political risks*.[107] Another group of risks – *infrastructure* – was added by the author of this study. As can be seen below, these risks and obstacles are somewhat interlinked and these divisions/titles can hardly be regarded as impermeable.

i Economic risks

As discussed above, the price of RE is falling. In some areas, it is even regarded as competitive on free market terms. There are, however, certain barriers that adversely affect the competitiveness of RE price and are therefore inhibiting investment and dissemination.

To begin with, the initial investment in fossil fuel-based facilities is much cheaper than that required for building new RE facilities.[108] It is true that once the initial investment has been made, the operating costs of RE generation are far cheaper (mostly due to the cost of fuels). Yet, the high cost of establishment implies also a need for substantial, long-term, upfront finance. The cost of borrowing could be significant enough to impact on investors' decisions and lead them to opt for the (initially) cheaper fossil fuels.

102 IEA, World Energy Investment Outlook (n 64) 32.

103 IEA, *Tracking Clean Energy Progress 2015* (IEA 2015) 12.

104 MIGA, *World Investment and Political Risk: World Investment Trends and Corporate Perspectives; the Political Risk Insurance Industry; Breach of Contract* (World Bank 2013).

105 MIGA is essentially an insurer; 'pre-claims events' are events in which insured projects 'have experienced some difficulty that had potential to result in a loss and claim by the investor'. See MIGA WIPR 2013 (n 104), 50.

106 MIGA WIPR 2013 (n 104) 50.

107 IPCC (n 99) 1225.

108 Tobias Schmidt 'Low-carbon investment risks and de-risking' (2014) 4 *Nature Climate Change* 237; Jing Hu et al. 'Barriers to investment in utility-scale variable renewable electricity (VRE) generation projects' (2018) 121 *Renewable Energy* 730, 735–736.

Also here, the case of Global South countries is more challenging: the price of borrowing is influenced *inter alia* by the risks that are embedded in the investment, and investments in the Global South are, as discussed below, often regarded as risky, and thus more expensive.[109]

Moreover, high initial investment cost implies that investors will expect higher prices in return for their investment. According to some, the resulting high prices can render RE unaffordable in developing countries.[110] Alternatively, the guaranteed prices offered by governments' support schemes may not be high enough to attract costly investment.[111] The high initial cost of RE such as solar is also mentioned as a barrier for the smaller home solar systems, where even in developed countries the price of installation is considered as high enough to deter home owners from investing in RE.[112] Where wealthy Western home owners are deterred by such prices, those residing in developing countries are likely to find these unaffordable.

The problem of cost recovery is even more challenging when it comes to rural areas, where energy poverty in the developing world is most abundant. Reliance on systems such as mini-grids and home grids, or, alternatively, the expansion of national grids to rural areas, is considered as more expensive than the supply of energy in highly dense areas that are well connected to national energy systems.[113] This implies higher electricity prices for residents in rural areas, and possibly also a need for some form of subsidies.[114]

Another element that affects the price of RE, or more accurately, the competitive position of RE investment, are fossil fuel subsidies. Estimates regarding the scope of fossil fuel subsidies vary to a great extent. The IEA, OECD and IRENA all estimate fossil fuel subsidies at between USD 300 and 380 billion.[115] The IMF's estimate is far higher: USD 5.3 trillion.[116] These variances are the result of different definitions as to what constitutes 'fossil

109 Schmidt (2014); Schwerhoff and Sy (n 48) 394. UNDP, *De-risking Renewable Energy Investment* (UNDP 2013), 32; IPCC (n 107) 1211; Janosch Ondraczek et al. 'WACC the dog: The effect of financing costs on the levelized cost of solar PV power' (2015) 75 *Renewable Energy* 888.

110 Ouedraogo (n 33) 3.

111 See for example the case of Zambia, in Oliver Johnson et al. (n 77) 20.

112 Kabir et al. (n 34) 897.

113 USAID 'When are renewable energy mini-grids more cost-effective than other options?' <https://www.usaid.gov/energy/mini-grids/economics/cost-effectiveness/>.

114 Rainer Quitzow et al., *The Future of Africa's Energy Supply: Potential and Development Options for Renewable Energy* (IASS Study 2016) 32.

115 According to the IEA fossil fuel consumption subsidies are increasing, totalling more than USD 300 billion per year in 2017. A REN21 report estimated this number at USD 370 billion. See IEA <https://www.iea.org/weo/energysubsidies/> and REN21 <http://www.ren21.net/gsr-2018/pages/highlights/highlights/#page-content>.

116 See table in Doug Koplow 'Defining and measuring fossil fuel subsidies' in Jakob Skovgaard and Harro van Asselt (eds.) *The Politics of Fossil Fuel Subsidies and Their Reform* (CUP 2018), 32; David Coady et al. 'How large are global fossil fuel subsidies?' (2017) 91 *World Development* 11.

fuel subsidies', the manner in which these were measured, and the number of countries that were surveyed.[117] Regardless of the difference in estimates, it is clear that the numbers are significant, and far higher than equivalent RE subsidies.[118] It is further clear that these subsidies are distorting energy markets, adversely affecting the competitiveness of RE,[119] and consequently also discouraging investors from investing in RE.[120]

Fossil fuel subsidies are also abundant in the developing world, and are likely to have a similar effect on investment in RE in these countries as well. The Overseas Development Institute estimates that in 2015 sub-Saharan countries provided fossil fuel subsidies totalling USD 26 billion.[121] Some of these countries, including *inter alia* Nigeria, Angola, Mozambique, Tanzania, Zambia, Zimbabwe and Côte d'Ivoire, provided more than USD 1 billion each in fossil fuel subsidies.[122]

ii Infrastructure

The lack of infrastructure is another barrier inhibiting private investment. Current levels of infrastructure-related investment are insufficient for meeting development goals. The 'infrastructure gap' has been identified by the international community on several occasions.[123] The Addis Ababa Action Agenda on Financing for Development has estimated the necessary finance at USD 1–1.5 trillion per year[124]:

> Establishing a new forum to bridge the infrastructure gap. Investing in sustainable and resilient infrastructure, including transport, energy, water and sanitation for all, is a pre-requisite for achieving many of our goals. To bridge the global infrastructure gap, including the $1 trillion to $1.5 trillion annual gap in developing countries, we will facilitate

117 Koplow (n 116) 32.
118 Koplow (n 116) 32.
119 Shelagh Whitley and Laurie van der Burg 'Reforming fossil fuel subsidies: The art of the possible' in Skovgaard and Harro van Asselt (eds.) (n 116) 52.
120 Richard Bridle and Lucy Kitson 'The impact of fossil-fuel subsidies on renewable electricity generation' (2014) IISD <https://www.iisd.org/sites/default/files/publications/impact-fossil-fuel-subsidies-renewable-electricity-generation.pdf> 13.
121 Shelagh Whitley and Laurie van der Burg 'Fossil fuel subsidy reform in sub-Saharan Africa: From rhetoric to reality' (2018) Working Paper <http://newclimateeconomy.report/workingpapers/wp-content/uploads/sites/5/2016/04/FFS-Reform-in-Africa_NCE-ODI_final.pdf> 7.
122 Whitley and van der Burg (n 121) 7.
123 See for example *G7 Ise-Shima Principles for Promoting Quality Infrastructure Investment* (2016) <https://www.mofa.go.jp/files/000160272.pdf>; *G20 Principles for Quality Infrastructure Investment* (2019) <https://www.mof.go.jp/english/international_policy/convention/g20/annex6_1.pdf>.
124 Addis Ababa Action Agenda of the Third International Conference on Financing for Development (UN 2015), para 14.

> development of sustainable, accessible and resilient quality infrastructure in developing countries through enhanced financial and technical support.

More recently, the OECD secretariat estimated that an investment of USD 6.3 trillion per year will be needed in order to meet the Sustainable Development Goals.[125] This figure increases to USD 6.9 trillion per year if the objectives of the Paris Agreement are to be met – an additional USD 2–3 trillion in comparison to current levels of investment.[126]

The lack of infrastructure in the developing world is striking, and the situation in Africa is especially alarming. According to Foster and Briceño-Garmendia '[o]n just about every measure of infrastructure coverage, African countries lag behind their peers in the developing world'.[127] Paul Collier adds in this respect: 'Africa needs far more infrastructure than its governments can afford to finance through tax or aid'.[128]

The lack of infrastructure is a major problem also in the more specific context of RE,[129] where vast investment in energy infrastructure is badly needed.[130] Authors have mentioned in this context technical problems such as old/inadequate grid infrastructure, transmission and distribution losses, unavailability of data, lack of backup/storage devices, lack of trained workforce and training facilities, lack of equipment and after-sale support services, problems with communications, and connectivity to the grid.[131]

Addressing the lack of infrastructure in the developing world through investment is not going to be easy. Investment in the Global South often implies higher risks, even in the case of traditional projects that are usually regarded as 'safe'. Paul Collier writes[132]:

> In OECD countries investment in electricity generation and water provision is classified as being in the utilities sector. Despite the prospects

125 OECD, The World Bank, UN Environment, *Financing Climate Futures: Rethinking Infrastructure* (OECD Publishing 2018) <https://www.oecd-ilibrary.org/environment/financing-climate-futures_9789264308114-en> 20.

126 OECD (n 125) 19–20.

127 Vivien Foster and Cecilia Briceño-Garmendia (eds.) *Africa's Infrastructure: A Time for Development* (The International Bank for Reconstruction and Development/The World Bank 2010) 2.

128 Paul Collier 'Attracting international private finance for African infrastructure' (2014) 1 *Journal of African Trade* 37.

129 Ouedraogo (n 33) 4.

130 According to some, in order to reach global access to energy in sub-Saharan Africa, a yearly investment of USD 55 billion is necessary in the next 15 years. Oliver Johnson et al. (n 77) 4.

131 See review in Sunil Luthra et al. 'Barriers to renewable/sustainable energy technologies adoption: Indian perspective (2015) 41 *Renewable and Sustainable Energy Reviews* 762, 765–768; IPCC (n 107) 1226; UNDP 2013 (n 109) 60.

132 Collier 2014 (n 128) 39.

> of slow growth in the OECD, it is rated as a safe but boring asset for a pension fund. However, in the process of moving from developed to African economies, safe boring assets morph into high risk, high required rate of return, ventures.

Collier adds in this respect other factors that can make investment in infrastructure controversial in their host states, and thus affect the political will to uphold commitments made to investors. These factors are especially relevant in the context of RE-related projects. For example, infrastructure investments could be opposed by local communities due to their 'intrusive' impact on the local environment.[133] Collier further mentions 'a reasonable presumption of corruption' which accompanies the operation of regulated industries in developing countries, as well as the political sensitivity of services that are perceived by the local population as elementary, or even as entitled government services.[134]

iii Regulatory infrastructure

Investments in RE are often long-term in nature and closely regulated. Investors, therefore, require clear and coherent legislation; they need to know what the rules and regulations currently are, and what they will be in the long-term future.[135] The rules that are relevant for investment in RE are diverse and include more than mere energy laws. Many other types of regulation, including on topics such as competition, land, intellectual property and contract enforcement, are also important for investors and will play a role in their decision to invest.[136]

The OECD has recognised the need for an existing, sound and comprehensive regulatory framework in the context of investment in RE. This organisation therefore issued a 'policy guidance' document, which is based on the assumption that a strong regulatory infrastructure will increase investment in RE in developing countries.[137] The World Bank (focussing on investment in the solar sector) also stated the need to ensure a 'full top-down normative coherence, from the constitutional level to that of local regulations' in order to attract private investors.[138] More specifically, this World Bank report mentioned a list of elements, including effective

133 Collier 2014 (n 128) 41.

134 Collier 2014 (n 128) 41.

135 For an international recognition of this necessity, see *G20 Principles for Quality Infrastructure Investment* (2019) (n 123), Principle 6.

136 OECD, *Policy Guidance for Investment in Clean Energy Infrastructure: Expanding Access to Clean energy for Green Growth and Development* (OECD 2015).

137 This document's objective is 'providing policy-makers with a set of issues to consider for strengthening their enabling environment for investment in the clean energy sector'. OECD, *Policy Guidance for Investment in Clean Energy Infrastructure* (n 136) 3.

138 World Bank, *A Sure Path to Sustainable Solar: Solar Deployment Guidelines* (n 98) 36.

dispute-resolution mechanisms, long-term land access and ownership, '[s]treamlined permitting processes; [a]dapted insurance regulations, [s]uitable foreign exchange regulations, [and] [c]lear tax provisions applicable to solar IPP-projects'.[139]

A comprehensive regulatory infrastructure is often missing in the developing world. Ouedraogo defines the RE regulatory infrastructure in African states as 'patchy and inconsistent'.[140] This author adds that in many countries relevant policies do not exist, and policies that are in place are not properly implemented and enforced.[141] Others have mentioned in this respect also the lengthy process of regulation and problems with issues such as the implementation of rules and secondary regulation, lacking institutional capacity (both in governance and in state-owned utilities), low levels of market liberalisation and more.[142]

iv Political risks

Another major barrier for RE-related investment (arguably the most significant of them all[143]) is related to stability and predictability.[144] Investors' demand for stability is affected by many variables, ranging from disturbing events such as wars and natural disasters to political stability, and the manner in which policies and laws are being applied and enforced. The UNEP's Green Economy Report summarises these factors:

> Country-specific institutional risks such as stability of the government, reliability of the legal system, transparency of business dealings, currency risks, and general instability due to wars, famine and strikes. For large-scale investments in a specific country, a long-term stable policy regime with a sound legal basis is needed.
>
> Political risk and regulatory risk such as unexpected changes in policy or uncertainty about the future direction of policy. Given the long pay-back periods, the contribution of policies to predictability, clarity

139 World Bank, *A Sure Path to Sustainable Solar: Solar Deployment Guidelines* (n 98) 36.

140 Ouedraogo (n 33) 4.

141 Ouedraogo (n 33) 4.

142 Quitzow et al. (n 114) 32. See a case study concerning delayed implementation in Oliver Johnson et al. (n 77) 19.

143 In an industry survey conducted by Gatzert and Kosub, political risks (including regulatory risks) were mentioned by most as a major barrier for investment in RE. See Gatzert and Kosub (n 98) 995.

144 For example, Steffen Brunner et al. 'Credible commitment in carbon policy' (2012) 12(2) *Climate Policy* 255. Some are distinguishing between 'political risks' and 'policy risks', the former addresses the more drastic risks (*inter alia* civil unrest) and the latter changes in regulations and policies (see Rogerio Studart and Kevin Gallagher 'Guaranteeing sustainable infrastructure' (2018) 255 *International Economics* 84, 86). In this study the two are addressed under the same title, as 'political risks'.

and long-term stability in the investment climate, are viewed as critical in being able to stimulate more investments[145];

The issue of policy uncertainties is especially important in the context of energy investment.[146] Any investment in the developing world implies a variety of risks. According to some, the most significant of these risks are political in nature.[147] Investors in the RE sector are often asked to make a very significant, one-off and irreversible investment, which will be recouped only over a significant number of years. Once the investment is made and the project has been delivered, the investor's ability to profit from the investment is entirely in the hands of the host state, which may, or may not, uphold its long-term commitments. It is understandable therefore that the value of stability and predictability is of immense importance for those considering investment in RE.

The effect that policy uncertainties have on climate-friendly investment was emphasised in a report compiled by Bloomberg New Energy Finance, in which the global 11% decline in private investment in clean energy that took place in 2012 was attributed mostly to policy uncertainties in markets such as the US, India, Italy and Spain.[148] This conclusion was repeated in a 2015 IEA report, in which it was stated that '[i]nvestments in renewable energy slowed in 2013, due to policy changes and uncertainties on renewable energy support in major economies such as Germany, Italy and France'.[149]

The EU Commission has also identified legislative long-term stability and reliability as a necessary means for reducing investment risks and encouraging investment in renewable sources of energy.[150] An OECD survey (presented in the chart below) supports these views, as among executives from Transnational Corporations (TNCs) participating in the survey 'clear, long term reduction targets' were the most agreed upon factor to be considered as 'very useful'.[151] The lack of clear renewable energy policies and targets is

145 UNEP, *Green Economy Report: Renewable Energy: Investing in Energy and Resource Efficiency* (UNEP 2011), 227.

146 Gianleo Frisari, Morgan Hervé-Mignucci, Valerio Micale and Federico Mazza 'Risk gaps: A map of risk mitigation instruments for clean investments' (2013) Climate Policy Initiative <http://climatepolicyinitiative.org/wp-content/uploads/2013/01/Risk-Gaps-A-Map-of-Risk-Mitigation-Instruments-for-Clean-Investments.pdf> 12.

147 Collier 2014 (n 128) 41.

148 Bloomberg New Energy Finance 'New investment in clean energy fell 11% in 2012' (14 January 2013), Bloomberg New Energy Finance, online: Bloomberg New Energy Finance <http://about.bnef.com/2013/01/14/new-investment-in-clean-energy-fell-11-in-2012-2/>.

149 International Energy Agency, *Track the Energy Transition: Where We Are, How We Got Here, and Where We Should Be* (IEA 2015), 6.

150 EU Commission, *Communication from the Commission: The Support of Electricity from Renewable Energy Sources* COM(2005)627, 7.12.2005, online: EU <http://eur-lex.europa.eu/LexUriServ/LexUriServ.do?uri=COM:2005:0627:FIN:EN:PDF> 16–17.

151 See in Céline Kauffmann and Cristina Tébar Less, *Transition to a Low-Carbon Economy: Public Goals and Corporate Practices* (OECD 2010) online: OECD <http://www.oecd.org/dataoecd/40/52/45513642.pdf> 81.

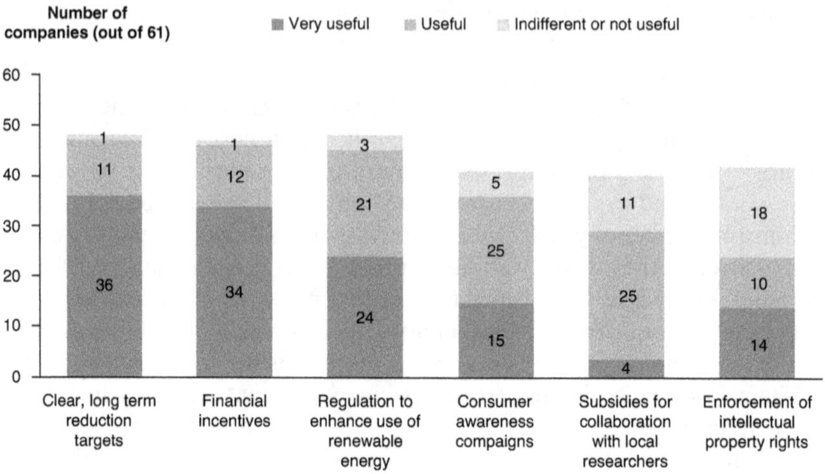

Source: OECD survey on business practices to reduce emissions.

especially noticeable in developing countries,[152] where the private sector's contribution is particularly important.

The effects of policy uncertainties can be simply explained by the fact that the lack of coherent policies is viewed as a potential business risk and therefore raises the costs of capital. High capital costs shift the balance in the decision-making process against investment in climate-friendly products. As explained below, at present, RE-related investment in the developing world is considered 'risky'. Without the certainty provided by stable regimes, the risk in making such an investment increases.[153] Long-termed, stable regulation provides in this respect transparency, planning security and stable business environment, all of which are supportive of climate-friendly investment as they reduce levels of risk.[154] This is especially important with respect to investment in new technologies, due to the long timeframes and large-scaled investment often necessary for the development and deployment of certain types of climate-friendly technologies.

For example, without long-term incentives such as feed-in tariff programmes or the setting of a minimum price for carbon, investment in

152 IISD 'Clean energy investment' (n 161) 18.

153 See also Richard G. Newell 'International climate technology strategies' (2008) Discussion Paper 2008–12 Harvard Project on International Climate Agreements, online: Harvard <http://belfercenter.ksg.harvard.edu/files/NewellWeb2.pdf> 12. About the timeframe necessary for the development of climate friendly technology, see UNEP SEF Alliance 'Public venture capital study' (2008) <https://energy-base.org/app/uploads/2020/03/2.Public-Venture-Capital-Study-2008.pdf> 12.

154 See also in UNEP, *Financing a Global Deal on Climate Change* (UNEP 2009) 8.

renewable energies may not be attractive.[155] This could be due to the ratio existing between the high risks involved in such an investment, on the one hand, and the low expected market-based returns, on the other. Long-term policies that will guarantee the prices of carbon, or the existence of feed-in tariffs for a fixed period of time, can solve this problem. The efficiency of public policies is indeed highly dependent on the provision of long-term stability.[156]

A study on venture-capital investment supports this conclusion and includes 'uncertainties of regulatory policies' and the 'absence of global climate change agreements' as part of the factors inhibiting venture-capital investors from investing in climate-friendly industry.[157] This was further supported by a study performed by Johnstone et al. which identifies a strong correlation between stable, predictable environmental policies and the level of environmental innovation (measured by patents' registration).[158] Stable public policies and laws are therefore necessary to support the private sector in this effort.[159]

Lastly, it has been argued that the lack of RE policies may signal to foreign actors that governments are not interested in applying climate change policies and therefore will not support related projects.[160] Furthermore, it has been argued by many that the current lack of climate change policies and regulation is problematic for private investors, also due to the uncertainties that such a situation creates. It is not clear, for example, what kind of measures will be taken by states, or when such will be established, if at all.[161] It was explained in this respect:

> the presence of a coherent national policy is better for the economy and their business than the status quo. Put another way, the absence

155 UNEP SEF Alliance 'Public venture capital study' (n 153) 12; Blyth et al. (n 162) 5766.

156 Catherine Mitchel, Dierk Bauknecht and Peter M. Connor 'Effectiveness through risk reduction: A comparison of the renewable obligation in England and Wales and the feed-in system in Germany' (2006) 34(3) *Energy Policy* 297, 305.

157 UNEP SEF Alliance 'Public venture capital study' (n 153) 11.

158 Nick Johnstone, Ivan Hascic and Margarita Kalamova 'Environmental policy uncertainty an innovation in environmental technologies' (2009) SSRN <http://papers.ssrn.com/sol3/papers.cfm?abstract_id=1524239>.

159 Brian Perusse, Matthew Riggins, Jeff Rodgers and Maury Zimring 'Melting down and scaling up: Stabilizing climate change by promoting private sector technology development' (2009) 26(4) *Review of Policy Research* 511; UN High Level Advisory Group on Climate Change Financing, Report of the Secretary-General's High Level Advisory Group on Climate Change Financing (2010) UN <https://www.cbd.int/financial/interdevinno/un-climate-report.pdf>.

160 IISD 'Clean energy investment' (n 161) 18.

161 Pew Centre on Global Climate Change 'In brief: The business case for climate legislation' (2010) online: Pew Centre <http://www.pewclimate.org/docUploads/business-case-for-climate-legislation-06-2010.pdf>; International Energy Agency, *Climate Policy Uncertainty and Investment Risk* (OECD/IEA 2007) online: IEA <http://www.iea.org/textbase/nppdf/free/2007/Climate_Policy_Uncertainty.pdf>; Aaron Cosbey, Jennifer Ellis, Mahnaz Malik and Howard Mann, *Clean Energy Investment: Project Synthesis Report* (IISD 2008), online: IISD <http://www.iisd.org/pdf/2008/cei_synthesis.pdf> 18.

> of such a policy creates uncertainty, which is a hindrance to sustained economic growth and an obstacle to the development of new markets and business opportunities.[162]

Delivering high levels of stability and predictability is a challenge that even developed countries such as the UK are not always meeting successfully.[163] One recent example concerns the changes made to the UK feed-in tariff programme. The impact of this change on the private sector can be understood from a statement made by the UK Renewable Energy Association following this change:

> The renewables industry is really not confident at all. These changes undermine confidence across all energy-related investments, even CCS [carbon capture and storage] and nuclear, as we need to be confident that the government will honour their commitments and not chop and change.[164]

A House of Commons Environmental Audit Committee's report added in this respect[165]:

> The UK has made impressive progress in reducing the carbon intensity of the UK power sector and driving investment in clean energy. But our predecessor committee warned in 2016 that sudden changes to policy had undermined investor confidence and could have an impact on the pipeline of projects in development. Recent figures for cash investment confirm that there has been a dramatic and worrying collapse in clean energy investment since 2015. Sudden policy changes have clearly played a significant part. [...] policy changes have dented investor confidence and curtailed the support available to new low carbon projects. The Government needs to restore confidence and provide a stable policy environment to deliver a pipeline of projects.

Spain provides a second example in which more drastic policy changes to long-term RE support schemes have occurred. A paper in which these

162 Pew Centre 'The business case for climate change' (n 161); William Blyth, Richard Bradley, Derek Bunn, Charlie Clarke, Tom Wilson and Ming Yang 'Investment risks under uncertain climate change policy' (2007) 35 *Energy Policy* 5766.

163 Gatzert and Kosub (n 98) 992–993.

164 Elena Ares, Oliver Hawkins and Paul Bolton 'Science and environmental section, 'feed-in tariffs: Solar PV'' (2012) online: UK Parliament <www.parliament.uk/briefing-papers/SN06112.pdf> 14.

165 The House of Commons Environmental Audit Committee, *Green Finance: Mobilising Investment in Clean Energy and Sustainable Development* (HC 2018) <https://publications.parliament.uk/pa/cm201719/cmselect/cmenvaud/617/617.pdf> 14.

changes are discussed describes severe consequences in the form of drastic reduction in solar installation.[166] The authors of this paper conclude: 'Indeed, the repeated changes and amendments had wider implications for renewable energy as a whole, damaging investor confidence in the reliability of Spanish policy frameworks'.[167]

> [...] [The] industry was generally critical of the changes because of their negative impact on the viability of the sector. The sector was particularly angered by retroactive changes to legislation. This changed the revenue streams upon which project development decisions had been made and permanently affected their confidence in the stability of Spain's regulatory environment. This latter issue caused two main problems: first, increased uncertainty for project developers; and second, greater uncertainty for lending institutions, which, in turn, equated to an increased cost of capital, and an associated rise in the cost of renewable energy.[168]

IV What? Explaining the objectives

As reviewed in the first part of this study, there is a need to increase RE-related investment in the developing world. The barriers on such investment, however, are significant, and the pace at which investment is presently being made is insufficient. Overcoming these barriers requires public intervention; notably public policies that will alleviate investors' concerns and resolve those market and policy obstacles that are currently stopping RE-related investments from materialising. The next question that will be addressed in this study is that of 'what?', or, more accurately, what should be the objective of these efforts?

The grand objective of public policies should, of course, be a significant increase in RE-related investment in the developing world. But as the integrationist spirit of the SDGs implies, merely achieving an increase in RE-related investment will not be enough. There are a good number of ancillary objectives that are implied by the achievement of the main objective, and must be realised in order to ensure that this objective will indeed result in *sustainable* development. For example, it is imperative to ensure that such investment will not result in human rights violation. It is also important to maintain environmental standards, as well as appropriate relationships and partnerships with indigenous communities. It is equally important to support local industry's growth and include an appropriate review mechanism to reflect fair returns. In short, the design of public policies must be

166 Pablo del Rio and Pere Mir-Artigues 'A cautionary tale: Spain's solar PV investment bubble' (2014) IISD/GSI <https://www.iisd.org/gsi/sites/default/files/rens_ct_spain.pdf>.
167 del Rio and Mir-Artigues (n 166) 2.
168 del Rio and Mir-Artigues (n 166) 21.

holistic, and integrationist in nature. This layer of integration is regarded as *integration of outcomes*; ensuring that the outcomes of investments will be sustainable, comprehensive and holistic, as much as possible.

The notion of 'integration' implies also a second layer; one that addresses not only the (wider) impact/outcome of investments, but also the necessity to attract investment in the first place. In other words, it is imperative to think of a variety of policies that are supportive of, or could potentially support, the scaling up of RE-related investment. States' policies, in this respect, must be wide and comprehensive. They should include a varied spectrum of legal tools that could, potentially, address the barriers discussed above, and ensure coherence and mutual-supportiveness. This layer is regarded as *integration of means*, i.e. ensuring that a variety of legal tools that are leading to increased investment are identified, designed in synergy and are mutually supportive.

The objective of public intervention, therefore, should be shaped in line with both types of integration; of *means*, and of *outcomes*. As stated above, designing integrationist policies is not an easy task. For a variety of reasons, states' past achievements in this respect have been modest, especially in the context of RE. The challenge of 'operationalising' the notion of integration is indeed both complex and enormous in scope. This study will not address this challenge in full; this study's aim is far more limited and specific in nature. It is aimed to support the design of holistic/integrationist policies from a very specific angle: that of international law home country measures. These two elements will be further explained in the next chapter of this study.

V Conclusion

This chapter places the foundation for this study. Admittedly, it is descriptive in nature: the reader may feel that the author has bombarded her with information which is not legal in nature and without much original analysis. This chapter, however, is an important first step. It explains why this study is necessary, and describes the barriers that all following chapters will now address.

The chapter ends by explaining the 'what' question – what kind of policies are needed in order to increase investment in the RE sector in the developing world. It hints at the importance of *integration* (of means, and outcomes). The next chapter will elaborate on this term, its role in achieving the SDGs, and the theoretical framework that is developed around it.

2 The tools

Home country measures and policy integration

I Introduction

Chapter 1 placed the factual foundations for this study by answering the 'why', 'who' and 'what' questions. The rest of this study will address the last, and most important question of all – the 'how' question – *how* can the international community achieve the grand objective of increasing investment in the renewable energy (RE) sector in the Global South?

There are, naturally, many answers for this 'how' question. This study will concentrate on the role of international law home country measures (HCMs), and the necessity to design these investment promotion instruments in an integrated, holistic manner. This focus requires a preliminary examination of two key terms that will guide the reader throughout this study: integration and HCMs. *Integration* is the principle that should guide the design of new policies. *HCMs,* are the type of policies that international law-makers should develop (or, at least, pay more attention to). The exploration of these two concepts will be the focus of this chapter.

The first part of this chapter will focus on policy integration. Authors have presented multiple theoretical frameworks and coined a variety of terms, all in essence reflecting one single imperative: policy integration. This chapter will start by presenting these theoretical framework(s) and explain the need to integrate policies from different areas. It will then concretise this discussion by explaining the necessity for policy integration in the specific context of this study: the efforts to promote RE investment in the developing world.

The second part of this chapter will be dedicated to the term 'international law HCMs'. The term 'international law HCMs' is, in essence, a combination of two elements: a unique legal system – international law – and a strategic choice of legal tools – home country measures. This part will provide some background information on the term HCMs; it will provide a definition for the term 'international law HCM'; it will explain the need to increase the use of international law HCMs, and the relevance of these measures in the specific context of promoting investment in the RE sector.

The two elements explained in this chapter – policy integration and international law HCMs – will be the focus of this study, and will guide the reader

DOI: 10.4324/9780429277207-3

throughout it. In the chapters to come, the reader will be exposed to the manner in which these terms were concretised in different areas in international law.

II How #1? Integration, SDG 17 and beyond

As stated above, the 'how' question essentially asks how should the international community address the challenge of increasing investment in the RE sector in the Global South? The first part of this study's answer is, *integration*. The challenge of increasing investment in the RE sector is a multi-layered endeavour. Investors' needs are varied, and so are the safeguards that must be placed in order to secure the long-term sustainability of RE projects. Moreover, many of the investment promotion instruments provided by international law are addressing only one element, or one investment barrier. But as explained in Chapter 1 of this study, there are numerous barriers standing in the way of foreign investors in the RE sector, and addressing only one of these will not necessarily tip the balance in favour of a decision to invest.

It seems clear therefore that any policy that will aspire to promote investment in the RE sector will have to rely on tools from a variety of areas. As reviewed below, there is a wide literature exploring concepts that are related to policy integration, including in the area of this study. The results of the examination provided in the chapters to come demonstrate that despite this growing awareness to integration, international laws and policies are still highly fragmented.

The following section provides a short review of the different concepts and terms that surround the notion of integration. It will then concretise this discussion and link it with the efforts to increase investment in the RE sector.

III Legal fragmentation, policy integration and beyond

As stated above, the efforts to promote investment in RE in the developing world must not be designed in isolation. A variety of obstacles and needs will have to be addressed, implying a *cross-disciplinary effort*, one that will incorporate elements from a myriad of legal areas. The greatest challenge embedded in cross-disciplinary efforts is related to coherence and coordination. This challenge goes beyond the need to ensure that the different policy-areas are all addressing a certain objective; it implies an effort to ensure that these parallel efforts are designed to achieve *synergies* with one another, and avoid conflicts.

The need for (and the lack of) policy coherence has been expressed, researched and debated under different guises, in a variety of academic fields. While there are certain differences between the different terms[1] and

1 See review in Frank Biermann et al. 'The fragmentation of global governance architectures: A framework for analysis' (2009) 9(4) *Global Environmental Politics* 14.

disciplines, the core issue remains essentially the same: international governance is complex, dense and too often lacking in coherence.

In legal scholarship, this problem is expressed mostly through the debate on the 'fragmentation of international law'. The phenomenon referred to as the 'fragmentation of international law' is rooted in a claim according to which different legal fields (e.g. international trade law, international investment law, international environmental law) have emerged in isolation; each designed by different law-makers/ministries, holding different perspectives, backgrounds and expertise. David Kennedy famously described the differences between these groups[2]:

> When we public international lawyers look out the window, we see a world of nation states and worry about war. We remember the great wars of the twentieth century. We were traumatized by the holocaust, fear totalitarianism and are averse to ideology. [...] Trade lawyers, by contrast, look out the window and see a world of buyers and sellers struggling to deal. Their trauma was the great depression.

Martti Koskenniemi wrote in this regard[3]:

> To be doing 'trade law' or 'human rights law', or 'environmental law' or 'European law' – as the representatives of those projects repeatedly tell us – is not just to operate some technical rules but to participate in a culture, to share preferences and inclinations shared with colleagues and institutions who identify themselves with that 'box'.

The fragmentation of international law, so it is claimed, is problematic for several reasons.[4] Briefly put, it is argued that the lack of coordination and coherence that define the international legal order could lead to conflicting instructions by different sources of international law, or to the issuing of conflicting decisions by different international tribunals.[5]

2 David Kennedy 'One, two, three many legal orders: Legal pluralism and the cosmopolitan dream' (2007) 31(3) *New York University Review of Law and Social Change* 641, 650.

3 Martti Koskenniemi 'International law: Between fragmentation and constitutionalism' Canberra (27 November 2006).

4 The reader should note that many disagree that indeed the 'fragmentation' is problematic. See for example the views of former ICJ Judges Bruno Simma and Christopher Greenwood. Bruno Simma 'Fragmentation in a positive light' (2004) 25(4) *Michigan Journal of International Law* 845; Christopher Greenwood 'Unity and diversity in international law' in Mads Andenas and Eirik Bjorge (eds.) *A Farewell to Fragmentation* (CUP 2015) 37.

5 See more in International Law Commission, *Fragmentation of International Law: Difficulties Arising from the Diversification and Expansion of International Law* (2006) A/CN.4/L.682; Joost Pauwelyn 'Bridging fragmentation and unity: International law as a

While international lawyers are concerned mostly with conflicting instructions/decisions, international relations (IR) authors have been addressing this challenge from a much broader perspective. Notably, IR authors have been studying the term 'institutional interaction', addressing the manner in which international institutions are affecting each other's operation and effectiveness.[6] IR authors are looking beyond conflicting legal instructions or judicial interpretations and are addressing much wider implications, namely, 'situations when the contents, operation or consequences of one institution (the recipient regime) are significantly affected by another (the tributary regime)'.[7] These impacts include a widespread notion of negative effects, as well as the concept of 'synergies', which is highly relevant for this current study.

The focus on potential synergetic and mutually supportive interactions that different policy-areas may have with one another is receiving increasing attention,[8] and more efforts are being directed towards the active design of mutually supportive policies. Academics and law-makers have addressed these deliberate efforts to design synergetic policies in numerous ways and under a variety of titles, including 'interplay management',[9] 'mainstreaming',[10] 'holistic governance', 'policy integration' and more.[11] While certain differences between these terms arguably exist,[12] they all nevertheless share the same nature and widely accepted objective[13]: the understanding that

universe of inter-connected islands' (2004) 25(4) *Michigan Journal of International Law* 903; Koskenniemi (n 3).

6 Sebastian Oberthür and Thomas Gehring, *Institutional Interaction in Global Environmental Governance* (MIT 2006); Thomas Gehring and Sebastian Oberthür 'The causal mechanisms of interaction between international institutions' (2009) 15(1) *European Journal of International Relations* 125.

7 Olav Schram Stokke 'The interplay of international regimes: Putting effectiveness theory to work' (2001) FNI Report 14/2001, at 2.

8 Ingrid Visseren-Hamakers 'Integrative environmental governance: Enhancing governance in the era of synergies' (2015) 14 *Current Opinion in Environmental Sustainability* 136.

9 Stokke (n 7) 11; Sebastian Oberthür 'Interplay management: Enhancing environmental policy integration among international institutions' (2009) 9 *International Environmental Agreements: Politics, Law and Economics* 371.

10 Joyeeta Gupta 'Mainstreaming climate change: A theoretical exploration' in Joyeeta Gupta and Nicolien van den Grijp (eds.) *Mainstreaming Climate Change in Development Cooperation* (CUP 2010) 67.

11 See full review in Jale Tosun and Achim Lang 'Policy integration: Mapping the different concept' (2017) 38 *Policy Studies* 553.

12 Tosun and Lang (n 11).

13 This concept was recognised in numerous international documents, agreements and declarations, including The recognition of this principle can be found in international documents such as the 1987 United Nations World Commission on Environment and Development's Brundtland Report, the 1992 Rio Declaration, the 1992 Climate Change Convention, the 2002 ILA New Delhi Declaration of Principles of International Law

sustainable development can be achieved only through the design of holistic, mutually supportive and mutually completing policies.

A The rise of policy coherence for sustainable development

The discussion on policy integration is as broad as it is important. The specific context of this study therefore requires a certain narrowing down. This study's focus is on the promotion of RE-related investment in the developing world. It is therefore required that a certain emphasis on development, and climate change, will be made. The integration of development-related considerations in wider policies is addressed through the terms 'Policy Coherence for Development' (PCD) and 'Policy Coherence for Sustainable Development' (PCSD). While some argue that the terms PCD and PCSD are essentially different,[14] it is difficult in practice to pinpoint such differences: notably, not many will challenge the view that *every* development must be sustainable. The two terms will therefore be regarded similarly in this study and will be used interchangeably throughout.

One widely[15] accepted definition describes PCD as 'taking account of the needs and interests of developing countries in the evolution of the global economy. It involves the systematic promotion of mutually reinforcing policy actions across government departments and agencies creating synergies towards achieving the agreed objectives'.[16] The reader will probably note the similarity between this definition and terms such as 'sustainable development' and 'mainstreaming'. Indeed these terms are fairly similar[17]; they all include the same relatively vague notion of integration and holistic policy-making.

As reviewed below, the terms PCD/PCSD were discussed and promoted mostly within organisations such as the Organisation for Economic

---

Relating to Sustainable Development, and the 2012 (Rio +20) United Nations Conference on Sustainable Development's outcome document ('The future we want').

14 Commission Staff Working Document '2019 EU report on policy coherence for development' (2019) SWD(2019) 20 final <https://ec.europa.eu/europeaid/sites/devco/files/swd_2019_20_pcdreport.pdf> 4; James Mackie et al. 'Policy coherence and the 2030 Agenda: Building on the PCD experience' (2017) ECDPM Discussion Paper No 10 <https://ecdpm.org/publications/policy-coherence-2030-agenda/>.

15 Carbone (n 38) 324. The reader should note that this definition was refined since by the OECD, although it is unclear whether new definitions indeed add something new. See a review of the evolution of the PCD's definition in OECD 'Policy coherence for inclusive and sustainable development' (2014) OECD and Post-2015 Reflections, Element 8, Paper 1 <https://www.oecd.org/dac/POST-2015%20PCD.pdf>.

16 OECD 'Policy coherence: Vital for global development' *Policy Brief* (July 2003) <http://www.oecd.org/gov/pcsd/20202515.pdf> 2.

17 Ulrike Zeigermann 'Governing sustainable development through 'Policy Coherence'? The production and circulation of knowledge in the EU and the OECD' (2018) 7(1) *European Journal of Sustainable Development* 133, 140.

Co-operation and Development (OECD) and the European Union (EU).[18] The OECD has been promoting PCD since the early 1990s, *inter alia* through the publication of studies, reports and guidelines on this topic.[19] The OECD has also issued several political declarations on PCD, prominent examples include the OECD 2002 Ministerial Declaration ('OECD Action for a Shared Development Agenda'),[20] and the 2008 Ministerial Declaration on Policy Coherent for Development, in which it was stated[21]:

> WE AGREE on the necessity of greater coherence and better co-ordination between the various international arrangements and institutions in order to help ensure that the benefits of globalisation are realised and broadly shared and to cope with the challenges it brings and maximise its benefits;

The terms PCD and PCSD have also found considerable traction within the EU. The notion of PCD was enshrined in the TFEU as an instruction: 'The Union shall take account of the objectives of development cooperation in the policies that it implements which are likely to affect developing countries'.[22] In 2010, the EU Parliament's Development Committee (DEVE) appointed a Standing Rapporteur on PCD, and several political statements have been made on this issue.[23] Following the adoption of the 2015 Sustainable Development Goals (SDGs), the EU's political commitment to PCD has dramatically increased. Notably, the EU Member States have responded to the adoption of the SDGs with their own vision on development (The New European Consensus on Development)[24] in which they commit to[25]:

> apply the principle of policy coherence for development (PCD), and will take into account the objectives of development cooperation in all external and internal policies which they implement and which are likely

18 See a review of the development of PCD within the EU and the OECD in Ulrike Zeigermann 'Governing sustainable development through "policy coherence"? The production and circulation of knowledge in the EU and the OECD' (2018) 7(1) *European Journal of Sustainable Development* 133.

19 See the OECD's 'hub' for PCSD <http://www.oecd.org/gov/pcsd/>.

20 See paragraph 5–6 in OECD, *OECD Action for a Shared Development Agenda* (2002) <http://www.oecd.org/newsroom/oecdcouncilatministeriallevel15-16may2002.htm>.

21 OECD, *Declaration on Policy Coherence for Development*, 4 June 2008 C/MIN(2008)2/FINAL, available online: <http://www.oecd.org/pcd/ministerialdeclarationonpolicycoherencefordevelopment.htm>.

22 See Art 208, Treaty on the Functioning of the European Union.

23 See for example the European Parliament resolution of 7 June 2016 on the EU 2015 Report on Policy Coherence for Development.

24 *The New European Consensus on Development* (n 54).

25 *The New European Consensus on Development* (n 54) para 10.

to affect developing countries. PCD is a fundamental part of the EU's contribution to achieving the SDGs.

At an international level, the most significant acknowledgement of the terms PCD/PCSD is found in the SDGs. These goals signify, according to some, a historical political consensus regarding the importance of synergies:

> [a] recognition of the fact that it is difficult, if not impossible, to achieve socio-economic gains if the environment, and threats such as climate change, are neglected; and that promoting synergies and mitigating or eliminating trade-offs among SDGs is essential for achieving this balance and for helping achieve the temperature targets of the Paris Agreement.[26]

Others have defined the adoption of the SDGs (and the Paris Agreement) as underpinning 'a new and transformational global agenda for sustainable development and climate action'.[27]

Notably, Goal 17s ('partnerships for the goals') targets include *inter alia* the title 'systemic issues', and the sub-title 'Policy and institutional coherence'. Under these titles, the following three targets are found[28]:

> **17.13** Enhance global macroeconomic stability, including through policy coordination and policy coherence
>
> **17.14** Enhance policy coherence for sustainable development
>
> **17.15** Respect each country's policy space and leadership to establish and implement policies for poverty eradication and sustainable development

Targets 17(13) and (especially) 17(14) reflect the different angles of the term PCD/PCSD. On the one hand, these targets provide a clear recognition of the importance of PCD and the role that PCD must play in future development efforts. On the other hand, these targets are extremely vague to the level that one must wonder what exactly they add to the already well-established

26 Luis Gomez-Echeverri 'Climate and development: Enhancing impact through stronger linkages in the implementation of the Paris Agreement and the Sustainable Development Goals (SDGs)' (2018) 376 *Philosophical Transactions of the Royal Society* <https://royalsocietypublishing.org/doi/pdf/10.1098/rsta.2016.0444>.

27 Karen Holm Olsen et al. 'Sustainability labelling as a tool for reporting the sustainable development impacts of climate actions relevant to Article 6 of the Paris Agreement' (2019) 19(2) *International Environmental Agreements: Politics, Law and Economics* 225.

28 UN 'Transforming the world: The 2030 agenda for sustainable development' (2015) A/RES/70/1.

term 'sustainable development'. The declaration attached to the SDGs does not add much clarity in this respect.[29] Target 17(15) similarly reiterates the oft-repeated disclaimer concerning states' sovereign right to essentially de-prioritise environmental protection in the face of economic/development considerations.

This notion of integration and coherence is expressed also in other instruments that are addressing more directly the topic of this study. The Addis Ababa Action Agenda on Financing for Development, which was concluded soon after the SDGs in order to support the Goals' implementation, is stating on the importance of policy coherence in the context of increased investment and finance[30]:

> We will respect each country's policy space and leadership to implement policies for poverty eradication and sustainable development, while remaining consistent with relevant international rules and commitments. At the same time, national development efforts need to be supported by an enabling international economic environment, including coherent and mutually supporting world trade, monetary and financial systems, and strengthened and enhanced global economic governance. [...]' 'We commit to pursuing policy coherence and an enabling environment for sustainable development at all levels and by all actors, and to reinvigorating the global partnership for sustainable development.

And[31]:

> Since Monterrey we have become increasingly aware of the need to take account of economic, social and environmental challenges, including the loss of biodiversity, natural disasters and climate change, and to enhance policy coherence across all three dimensions of sustainable development. We will take measures to improve and enhance global economic governance and to arrive at a stronger, more coherent and more inclusive and representative international architecture for sustainable development, while respecting the mandates of respective organizations. We recognize the importance of policy coherence for sustainable development and we call upon countries to assess the impact of their policies on sustainable development.

29 See for example paragraph 63, which essentially repeats existing consensuses and does not add any additional clarity.
30 Addis Ababa Action Agenda of the Third International Conference on Financing for Development (UN 2015), para 9.
31 Addis Ababa Action Agenda of the Third International Conference on Financing for Development (UN 2015), para 103.

B *Environmental policy integration (EPI) and climate policy integration (CPI): in search of the optimal compromise*

The efforts to implement PCD are derived from those wishing to promote *development* and integrate/streamline developing countries' needs and interests into wider policies. Those who wish to promote *environmental protection*, on the other hand, are attempting to use the same integrationist tool, but from a different angle – they wish to integrate/streamline environmental considerations into wider policies, including policies that are related to development. The context of this study – investment in *renewable energy* in the *Global South* – requires attention for both groups and terms, as well as to the relations between these objectives.

The conceptual frameworks pursued by those wishing to promote environmental/climate goals are those of Environmental Policy Integration (EPI) and Climate Policy Integration (CPI).[32] The exact objective of these two terms has been a matter for debate.[33] Notably, it is uncertain whether the two represent an aspiration to *prioritise* environmental/climate goals over others, or simply to follow the 'mother concept' – sustainable development[34] – and focus on the balancing of the economic, environmental and social spheres. The adoption of the SDGs and their prominence in global governance suggest that the latter interpretation will be regarded as more acceptable.[35] This interpretation also ensures greater coherence with the above-discussed notion of PCD/PCSD.

The need for a holistic approach is clear and well agreed on. At the same time, it is crucial to add that while fully balanced and integrated results are desirable, certain prioritisation is often inevitable as 'win-win' scenarios are not easy to attain. The difficulty in attaining a truly balanced, sustainable development-compatible result is evident if one is to consider that environmental policies on their own may be conflicting and lacking in coherence. In fact, the terms EPI and CPI are conflicting at times, as climate mitigation and adaptation projects could result in environmental harm, and vice versa.[36] For example, investment in projects such as windfarms, hydroelectric facilities or nuclear power plants may be compatible with climate change goals, but harmful for bio-diversity and habitats. Moreover, climate

32 Camilla Adelle and Duncan Russel 'Climate policy integration: A case of déjà vu?' (2013) 23 *Environmental Policy and Governance* 1.

33 Andrew Jordan and Andrea Lenschow 'Environmental policy integration: A state of the art review' (2010) 20 *Environmental Policy and Governance* 147.

34 Jordan and Lenschow (n 33) 148.

35 See more specifically targets 17(13)–17(15) of the SDGs (see more on this topic below), United Nations General Assembly, *Transforming our World: The 2030 Agenda for Sustainable Development* (2015) A/RES/70/1. See also Nilsson and Persson (n 47) 37.

36 Katharina Rietig 'Sustainable climate policy integration in the European Union' (2013) 23 *Environmental Policy and Governance* 297.

mitigation projects could potentially affect development-related objectives, including the most basic rights of humans and communities. Climate policies may resultantly impact on displacement, the re-allocation of resources, controversial land-use and violation of human rights.[37]

It is important at this early stage to remind the reader that trade-offs are an essential element in the design of EPI/CPI, as indeed is the case with almost any attempt to achieve policy integration and coherence.[38] For example, the re-direction of Official Development Assistance (ODA) towards climate mitigation goals essentially implies that fewer resources will be available for goals such as education or water sanitation.[39] Equally, the liberalisation of international trade in environmental goods and services will mean less protection for domestic infant/less competitive industries, and the protection of foreign investors in the RE sector will inevitably restrict host states' regulatory flexibility. These are all examples of measures that are recommended in this study, but will certainly come at a cost. There is no way around such trade-offs, and any search for a puritan, fully balanced 'win-win' solution will be both futile and wasteful.

This is not to argue that efforts to explore and minimise the adverse effects of CPI should not be made. The opposite is true; to claim that any efforts to improve policy coherence will have to address conflicting objectives and outcomes will be stating the obvious. The point is however, that no solution will be popular with all, and trade-offs are a necessity that must be acknowledged and accepted.

C The practice of PCD, environmental and climate integration: easier said than done

The understanding that climate change is an extremely multi-faceted, cross-sectorial and cross-disciplinary issue ('a super wicked problem'[40]) is hardly ground-breaking. The same is true also for the need to integrate

37 There is wide literature on the impact of CDM projects on human rights. See for example Mary Finley-Brook and Curtis Thomas 'Renewable energy and human rights violations: Illustrative cases from indigenous territories in Panama' (2011) 101(4) *Annals of the Association of American Geographers* 863; Naomi Roht-Arriaza '"First, do no harm": Human rights and efforts to combat climate change' (2009) 38 *Georgia Journal of International and Comparative Law* 593; Frances Seymour 'Forests, climate change and human rights: Managing risks and trade-offs' in Stephen Humphrey (ed.) *Human Rights and Climate Change* (CUP 2010) 207.

38 Maurizio Carbone 'Mission impossible: The European Union and policy coherence for development' (2008) 30(3) *European Integration* 323, 326.

39 Nevertheless, an OECD report has questioned whether such a diversion of funds indeed occurs in reality. See OECD, *Climate Finance Provided and Mobilised by Developed Countries in 2013–17* (OECD 2019) 28.

40 Richard Lazarus 'Super wicked problems and climate change: Restraining the present to liberate the future' (2009) 94 *Cornell Law Review* 1153.

climate change considerations into policies from all spheres of life. The necessity to integrate climate policies was recognised in international documents, including the text of the UNFCCC[41] and following decisions/agreements.[42] While the idea of integration is well-accepted, the deployment of this concept into practice is a different and somewhat more ambiguous matter.

At an international and multilateral level, practical instructions on how to implement the notion of integration are almost always absent. At most, one can expect to find only vague declarations in preambles concerning the importance of integration.[43] The causes for this lack of practical instruction are often the difficulties involved in dealing with details and differing circumstances at multilateral levels, as well as the political objective of reaching consensus in major international decisions and the consequent reluctance to address difficult trade-offs. Nilsson et al have commented on the challenge of delivering integration in the context of the SDGs[44]:

> Implicit in the SDG logic is that the goals depend on each other – but no one has specified exactly how. International negotiations gloss over tricky trade-offs. Still, balancing interests and priorities is what policy-makers do – and the need will surface when the goals are being implemented. If countries ignore the overlaps and simply start trying to tick off targets one by one, they risk perverse outcomes.

Some efforts are nevertheless taking place. A large number of states have committed to mainstream climate change objectives in their development policies,[45] and international organisations are attempting to facilitate this

41 See Art 3(4) of the UNFCCC.

42 See for example Art 1(b) of UNFCCC, Decision 5/CP.17, FCCC/CP/2011/9/Add.1 (2011), according to which National Adaptation Plans' (NAP) objective is 'to facilitate the integration of climate change adaptation, in a coherent manner, into relevant new and existing policies, programmes and activities, in particular development planning processes and strategies, within all relevant sectors and at different levels, as appropriate'.

43 The most well-known example being the preamble to the WTO Agreement, in which it is stated *inter alia* that trade liberalisation should be achieved

> in accordance with the objective of sustainable development, seeking both to protect and preserve the environment and to enhance the means for doing so in a manner consistent with their respective needs and concerns at different levels of economic development.

Agreement Establishing the World Trade Organisation (1994). See review in the author's previous work in Kent (n 62); Nilsson and Persson (n 47) 36.

44 Mans Nilsson et al. 'Policy: Map the interactions between Sustainable Development Goals' (2016) 534 *Nature* 320.

45 According to Kathleen Mogelgaard et al., 30 developing countries have addressed in their NDC efforts to mainstream climate adaptation, as well as a large group of developed countries. Kathleen Mogelgaard et al. 'From Planning to Action: Mainstreaming Climate Change Adaptation into Development' (2018) World Resources Institute Working Paper, < https://www.wri.org/ publication/climate-planning-to-action > 9.

effort through the publication of guidelines.[46] The translation of these commitments into successful, practical working policies however, remains challenging and levels of implementation are currently limited.[47] The reasons for this state of affairs are numerous, and include the proliferation of policy objectives and actors, conflicting interests,[48] competition over dwindling resources and a fragmented institutional framework.[49]

The difficulties in achieving coherence are related also to what was described by the UNCTAD Secretariat as 'administrative complexity'.[50] This term reflects a situation in which different, administratively isolated groups (e.g. different ministries[51]) are negotiating different international legal arrangements, and despite obvious links and dependencies, very little coordination and communication between these groups takes place. While this state of affairs is especially difficult for developing countries (which 'may find this situation particularly difficult to navigate'[52]), it is certainly also affecting rich nations, with sophisticated and well-funded civil-services.

Other difficulties in delivering policy coherence are resulting from the hard trade-offs that are inherent in the process of integration. As already discussed, the prioritisation of one element often implies the de-prioritisation of another, and one must accept that in some cases, a 'win-win' solution is simply not on the table. One relevant example is related to the use of ODA for the promotion of investment in the RE sector in the developing world. As will be discussed in Chapter 6 of this study, as part of their efforts to support SDG 13 (climate action), donor states tend to direct their climate-related ODA funds towards destinations that are 'cost-effective', i.e. will lead to emission reductions with minimal risks and costs. Consequently, most climate-related ODA funds are directed towards middle-income countries, where investment conditions and risks are preferable over those existing in

46 See for example UNDP/UNEP, *Mainstreaming Climate Change Adaptation into Development Planning: A Guide for Practitioners* (UNDP/UNEP, 2011); OECD, *Integrating Climate Change Adaptation into Development Cooperation: Policy Guidance* (OECD 2009).

47 Mogelgaard et al. (n 45) 9; Mans Nilsson and Asa Persson 'Lessons from environmental policy integration for the implementation of the 2030 Agenda' (2017) 78 *Environmental Science and Policy* 36.

48 Barder et al. (n 59) 847; Anna Pegels et al. 'Politics of green energy policy' (2018) 27(1) *Journal of Environment Development* 26, 30.

49 See review in Carbone (n 38) 326–327. On institutional fragmentation, see also Avidan Kent 'Implementing the principle of policy integration: Institutional interplay and the role of international organizations' (2014) 14(3) *International Environmental Agreements: Politics, Law and Economics* 203.

50 UNCTAD, *World Investment Report 2018* (UNCTAD 2018) 112.

51 These groups are isolated in many ways, beyond their administrative divisions. See for example Koskenniemi and Kennedy's remarks above concerning the differences in perspectives, expertise, background, etc.

52 UNCTAD WIR 2018 (n 50) 113.

Least Developed Countries.[53] The result of this trend is the prioritisation of SDG 13 over SDG 7 (rather than the integration of the two), as relevant ODA support is not reaching the destinations in which energy poverty is most prominent.

The European Union's experience on this matter is revealing. The EU's 1998 'Cardiff Process' ignited a long-lasting (ongoing) attempt to integrate environmental concerns in all areas of the EU's operation. This aspiration was manifested *inter alia* in (now) Article 11 TFEU, which states that '[e]nvironmental protection requirements must be integrated into the definition and implementation of the Union's policies and activities, in particular with a view to promoting sustainable development'. The EU added to this ambitious task layers of strategies, including its Sustainable Development Strategy and Europe 2020. In its (2017) official response to the 2030 Agenda for Sustainable Development,[54] the emphasis on integration and coherence is especially noticeable. It is stated *inter alia*[55]:

> When planning and implementing development cooperation, the EU and its Member States will pay particular attention to such interlinkages and to integrated actions that can create co-benefits and meet multiple objectives in a coherent way.

Opinions on whether the implementation of environmental integration within the EU was successful or not are divided.[56] The EU Commission's website and publications often create the impression that integration is happening in practice.[57] A careful reading of studies (even those commissioned by the Commission) suggest that the picture is somewhat more ambiguous,[58] where some even go as far as referring to the EU's efforts to implement terms such as PCD as 'lip-service to enhancing PCD without making

53 Homi Kharas 'Aid and climate finance' (2015) *Brookings* <https://www.brookings.edu/wp-content/uploads/2016/07/aid-climate-finance-kharas-1.pdf> 26; Giorgio Gualberti and Lorcan Lyons, *Missing the Mark: Gaps and Lags in Disbursement of Development Finance for Energy Access* (Sustainable Energy for all 2017) <https://www.afdb.org/fileadmin/uploads/afdb/Documents/Publications/Missing_the_Mark_Gaps_and_Lags_in_Disbursement_of_Development_Finance_for_Energy_Access.pdf> 8.

54 EU, *The New European Consensus on Development 'Our World, Our Dignity, Our Future'* <https://ec.europa.eu/europeaid/sites/devco/files/european-consensus-on-evelopment-20170602_en.pdf>.

55 EU, *Our World, Our Dignity, Our Future* (n 54) para 19.

56 See review in Adelle and Russell (n 32).

57 See for example 2019 EU report on policy coherence for development (n 14) 10–11.

58 Examples of studies in which difficulties to apply EPI/CPI are described include the European Commission 'Study on the integration of environmental concerns in the Cohesion Policy funds (ERDF, ESF, CF)' (2017) <http://ec.europa.eu/environment/integration/pdf/enea/Cohesion%20Pol_COWI-Milieu_December2017.pdf>; Duncan Russell et al. 'Understanding policy integration in the EU – Insights from a multi-level lens on climate adaptation and the EU's coastal and marine policy' (2018) 82 *Environmental Sciences an*

real efforts for systematically analyzing the effects of EU policies in developing countries'.[59]

With respect to the more specific topic of this research, it has been commented by authors that '[s]urprisingly little is known about the effectiveness of these EPI instruments'[60] and that 'the inclusion of climate concerns in EU external relations is still largely uncharted territory'.[61] An initial study conducted by the author on the integration of climate-related goals and tools in a variety of EU international treaties suggests that indeed, despite clear political statements, the presence of substantive legal instructions on the promotion of climate-related goals in non-environmental treaties is very rare.[62]

D Concretising the discussion: linking development, investment, and climate change

As stated above, terms such as EPI/CPI, and CPD/CPSD follow the lines of previous 'buzz-words' in the world of international law and policy, notably 'sustainable development' and 'green economy'. Delving into these concepts may seem somewhat repetitive and even redundant as all of these terms essentially represent one very simple idea – policies must not be made in isolation.

A useful step in this respect would be to move away from the conceptual discussion and address more concrete cases of policy-areas in which the implementation of EPI/CPI, PCD, etc., is relevant and necessary. This study intends to do exactly this; it will focus on very specific elements that are

---

Policy 44; Frederik De Roeck et al. 'Mainstreaming climate change adaptation into the European Union's development assistance' (2018) 81 *Environmental Sciences and Policy* 36.

59 Zeigermann explains:

> Between 2007 and 2015 analyses have mainly dealt with evaluating bureaucratic arrangements, technical and managerial processes within the EU. Therefore, authors have criticized that the EU and its member states are only paying lips-service to enhancing PCD without making real efforts for systematically analyzing the effects of EU policies in developing countries.

Zeigermann (n 14) 136. Carbone and Keijzer as well are describing the EU's efforts as 'lip service', adding that the 'results achieved have by no means been commensurate with the expectations raised by the various reforms launched over the years'. Maurizio Carbone and Niels Keijzer 'The European Union and policy coherence for development: Reforms, results, resistance' (2016) 28 *European Journal of Development Research* 30, 31; see also evaluation by Owen Barder et al. 'Europe beyond aid: Assessing European countries' individual and collective commitment to development' (2013) 25 *Journal of International Development* 832, 847.

60 Jordan and Lenschow (n 33) 155.

61 De Roeck et al. (n 58) 36.

62 Avidan Kent 'Policy coherence and the promotion of foreign direct investment in the renewable energy sector: Lessons from Europe' in Lisa Sachs, Lise Johnson and Jesse Coleman (eds.) *2017 Yearbook on International Investment Law and Policy* (OUP 2018) 22.

related to international law instruments and the manner in which a variety of policies are addressing investment in renewable energy-related projects in the developing world.

The topic of this study – RE-related investment in the Global South – fits within more than one of the above frameworks. It can be seen as an implementation of CPI as the outcome of RE-related projects is the reduction in GHG emissions. This topic can also be seen as an implementation of PCD/PCSD, as RE-related projects are supporting developing countries' need for energy and often come with additional positive side effects, such as technology transfer, jobs and a healthier environment. This study could also be regarded as an implementation of the (UNEP-championed) term 'green economy', or the wider 'mother concept' – sustainable development. There is no need to decide on the 'right' conceptual framework; these terms are not mutually exclusive and, once again, at the heart of these efforts sits the same idea – integration.

The literature on the integration of climate change and development policies is fairly wide. During the 1990s–2000s authors explored the need to connect these two spheres and find mutually supportive elements.[63] In following years, authors began researching more specific interactions within the climate-development nexus, such as in the context of specific sectors (e.g. land use,[64] agriculture,[65] health, migration,[66] and more[67]) or regions.[68] It is widely agreed today that there is a 'strong justification for the urgent need to carefully examine the linkages between climate and development to exploit the synergies and avoid or mitigate the negative trade-offs'.[69]

In the specific context of investment in renewable energies, Newell et al have argued already in 2009 that '[w]e know little, however, about the governance of CD [clean development, AK]: which features of these actors, institutions, and policy-making processes are resulting in effective outcomes in terms of

63 See for example Richard Klein 'Integrating mitigation and adaptation into climate and development policy: Three research questions' (2005) 8 *Environmental Science & Policy* 579; John Robinson and Deborah Herbert 'Integrating climate change and sustainable development' (2001) 1(2) *International Journal of Global Environmental Issues* 130; Marcel Kok et al. 'Integrating development and climate policies: National and international benefits' 8 (2008) *Climate Policy* 103; Lisa Schipper 'Climate change adaptation and development: Exploring the linkages' (2007) Tyndall Centre Working Paper No 107.

64 See for example Monica Di Gregorio 'Climate policy integration in the land use sector: Mitigation, adaptation and sustainable development linkages' (2017) 67 *Environmental Science & Policy* 35.

65 See for example Leslie Lipper et al. 'Climate-smart agriculture for food security' (2014) 4 *Nature Climate Change* 1068.

66 Avidan Kent and Simon Behrman, *Facilitating the Resettlement and Rights of Climate Refugees* (Routledge 2018).

67 See review in Ch. 5 of IPCC, *Global Warming of 1.5°C* (2018) <https://www.ipcc.ch/sr15/>.

68 See for example Lindsay Stringer et al. 'Advancing climate compatible development: Lessons from Southern Africa' (2014) 14(2) *Regional Environmental Change* 713.

69 Gomez-Echeverri (n 26) 3.

climate action and developmental benefits, which are not and why'.[70] They add to this 'unknown' also the fact that in the context of RE-related investment, '[t]he scale of the governance challenge, requiring interventions across levels by a multitude of actors, is quite possibly unprecedented'.[71]

The UNCTAD Secretariat has addressed the need to 'synergise' investment treaties with climate change policies already in this organisation's 2010 World Investment Report (WIR). The 2010 WIR calls for the integration (or 'mainstreaming') of investment policies into climate change policies,[72] stating that[73]:

> Securing IIAs' [International Investment Agreements, AK] contribution to climate change mitigation entails introducing climate-friendly provisions into future IIAs. A multilateral understanding is needed to ensure the coherence of existing IIAs with international and national policy developments related to climate change.

The 2010 WIR however is focussed almost exclusively on the improvement of investment treaties, rather than the exploration of wider mutually supportive links with other instruments.[74] The issue of policy coherence continued to feature in the work of the UNCTAD ever since,[75] although it is not clear whether the UNCTAD's Secretariat's view of terms such as 'coherence' and 'synergies' has expanded. The 2018 UNCTAD WIR addresses two levels of 'coherence'. The first is between international and national policies; an important topic on its own, but one that is beyond the scope of this study. The second level mentions improvements in coherence and synergies between different 'bodies of international law' as an essential part for reforming the field of investment law. Surprisingly, however, the narrow view of 'coherence' and 'synergies' continues to dominate also the 2018 UNCTAD WIR and, indeed, the WIR discussion's point of departure remains the narrow 'fragmentation of international law', which is focussed almost exclusively on legal conflicts and the interpretation of treaties. The UNCTAD WIR's 'policy options' on improving policy coherence in IIAs state in this respect: 'Addressing this relationship in IIA reform can help

70 Peter Newell, Nicky Jenner and Lucy Baker 'Governing clean development: A framework for analysis' (2009) 27(6) *Development Policy Review* 717, 719.

71 Newell et al. (n 70) 720.

72 UNCTAD, *2010 World Investment Report* (UNCTAD 2010) 136–138.

73 UNCTAD 2010 WIR 136.

74 For example, the WIR 2010 is discussing the manner in which IIAs can allow more regulatory flexibility, or the 'holistic' interpretation of IIAs by investment tribunals, and the prevention of legal conflicts. There is no discussion of other legal mechanisms from other fields (e.g. trade, insurance, environmental law) that also promote investment and could be *linked* or *supplemented* by IIAs.

75 UNCTAD, *2018 World Investment Report* (UNCTAD 2018) 104.

avoid conflicts and provide arbitral tribunals with guidance on how to interpret such interaction'.[76]

The UNCTAD WIR's narrow 'legal' focus is perhaps understandable, given the legal nature of the WIR and its focus on IIAs. The UNCTAD's recommendations in this respect are certainly useful and indeed will be discussed in the relevant section of this study. It should also be noted that the wider understanding of 'synergies' is not entirely absent from the 2018 UNCTAD WIR, and that it does (albeit briefly) recommend the exploration of wider synergies (referred to as 'cross references'), such as between investment agreements and the Agreement of the International Monetary Fund.[77] Nevertheless, overall, the UNCTAD's view seems too narrow and much of the potential of wider interactions[78] seems to be neglected.

The OECD is another institution that works to promote investment in renewable energy. Importantly for this study, this organisation published a Policy Guidance for Investment in Clean Energy Infrastructure[79]; an extensive document that covers a comprehensive number of fields, including investment, finance, competition, land, IP, and more. Even if some legal angles are mentioned only in passing (e.g. the World Trade Organisation) or not at all (e.g. environmental law, human rights law), this document represents an ambitious and comprehensive attempt to address investment in RE from the perspective of policy coherence.

Regarding the specific implementation of policy coherence, the OECD Policy Guidance provides a few concrete examples of mechanisms through which specific coordination is achieved (e.g. the relations between competition policies and energy regulators[80]). There is also an elaborated part on 'governance' in which the importance of coordination (on municipal, national and, to a lesser extent, also regional levels) is stressed.

While this report is certainly important, it is focussed almost exclusively on national laws. With very few exceptions,[81] international law is mostly ignored. The role of national law and national authorities is obviously important, and there is no doubt that any attempt to encourage international RE-related investment should begin there. The focus of this study, however, is on international law, and the OECD Policy Guidance is lacking in this respect.

76 UNCTAD, 2018 WIR (n 75) 111.

77 UNCTAD, 2018 WIR (n 75) 114.

78 As understood in disciplines such as IR and implied by terms such as PCD, EPI and 'mainstreaming' and 'sustainable development'.

79 OECD, *Policy Guidance for Investment in Clean Energy Infrastructure* (OECD 2013) <https://www.oecd.org/daf/inv/investment-policy/CleanEnergyInfrastructure.pdf>.

80 OECD, Policy Guidance (n 79) 41–42.

81 The OECD Policy Guidance includes a short of discussion of WTO Law and regional integration.

The OECD has published many other relevant documents on investment and climate change, some of which have addressed PCD in the context of RE-related investment, albeit in a more limited manner.[82] The OECD's Better Policies for Sustainable Development document, for example,[83] provides a brief review of investment in RE. The coverage of synergetic, mutually supportive policies in this document is relatively limited, especially in the context of international law.[84]

The one *international* area in which the OECD work on PCD and RE-related investment does often address mutually supportive, synergetic policies, relates to the use of development assistance. The OECD's 2016 Better Policies for Sustainable Development document states *inter alia* that 'official development assistance (ODA) remains essential in creating an enabling global environment for green growth while supporting specific measures in developing countries'.[85] The OECD mentions in this respect the targeting of ODA for the promotion of private investment, for capacity building and the facilitation of international trade in environmental goods and services.[86] A different (2012) OECD report states in this respect[87]:

> It is clear that if the opportunities offered by the transition to a green economy are to be realised, all nations need to be brought on board – including the refocusing of official development assistance (ODA) to support green economy and growth in developing countries. For the poorer countries, however, big obstacles remain; these range from complex issues – such as improved governance – to simple realities, such as the currency exchange risks that dampen the enthusiasm of investors, no matter how vast the opportunity offered by a wind or solar resource project may be. If carefully and strategically deployed, ODA could play a catalytic role in overcoming some of these barriers, as could implementing and financially supporting a green climate fund.

E Integration: the challenge ahead

While wide recognition of the value of PCD seems to exist, it is clear from the above that the challenge ahead mainly involves the concretising of terms and ideas on integration into specific, actual laws and policies. The concept

82 See for example OECD, *Aligning Policies for a Low Carbon Economy* (OECD 2015).

83 OECD, *Better Policies for Sustainable Development 2016: A New Framework for Policy Coherence* (OECD 2016); Dirk Röttgers 'The government role in mobilising investment and innovation in renewable energy' (2017) OECD Investment Insights <https://www.oecd.org/investment/The-government-role-in-enabling-investment-and-innovation-in-renewable-energy.pdf>.

84 OECD, *Better Policies for Sustainable Development* (n 83) 199.

85 OECD, *Better Policies for Sustainable Development* (n 83) 205.

86 OECD, *Better Policies for Sustainable Development* (n 83) 205.

87 OECD, *Development Co-operation Report 2012: Lessons in Linking Sustainability and Development* (OECD Publishing 2012) http://dx.doi.org/10.1787/dcr-2012-en at169.

of integration, in its many forms and guises, is 'easier said than done'. The devil is always in the detail; real synergies require complex and technical examinations and a good amount of cross-disciplinary and cross-institutional expertise. Where administrative units are operating in isolation, achieving integration will be extremely hard to accomplish.

This study will take several steps that, hopefully, will assist states in addressing this challenge. It will review specific areas of international law, seek potential synergies and provide more concrete recommendations as to what kind of action should be taken in more specific areas of law.

For reasons of scope and time, this study's recommendations will be limited. The reader should be aware that deeper examinations of synergies will need to be conducted by states if the challenge imposed by SDG 7 is to be met. Moreover, as stated by the author in the past,[88] the challenge of achieving integration is beyond this study's recommendations: This challenge is not only legal, but (and perhaps mostly) administrative in nature. The institutional fragmentation (i.e. in ministries, IGOs and other administrative bodies that support the law-making process) needs to be addressed in order to facilitate the creation of truly integrated policies.

IV How #2? Home country measures

A Definitions

This chapter aims to address the 'how' question – how should the international community address the challenge of increasing investment in the RE sector in the developing world? The first part of this study addressed the first part of the answer: *integration*. The second part of the answer offered by this study concerns the increased use of international law Home Country Measures.

Home country measures (HCMs) are measures that investors' home states (mostly investment exporting states) are taking in order to support and promote their nationals' investment in foreign markets. More elaborately, the UNCTAD Secretariat defined HCMs as[89]:

> [P]olicy measures taken by the home countries of firms that choose to invest abroad designed to encourage FDI flows to other countries. Their formulation and application may involve both home and host country Government and private sector organizations.
>
> HCMs exist at the national, regional and multilateral levels and involve a broad variety of measures, ranging from information provision, technical assistance and capacity-building, to financial, fiscal and insurance measures, investment-related trade measures, and measures

88 Kent (n 49).
89 UNCTAD, *Home Country Measures* (UNCTAD 2001) 66.

> related to the transfer of technology. Given this variety, HCMs have to be adaptable and flexible, since 'no one size fits all'.

Elsewhere, HCMs were defined as[90]:

> [T]he granting of specific advantages by the home country government (or one of its public institutions) in connection with the establishment, acquisition and expansion of an investment by a home country firm in a foreign economy [FN omitted]. They are meant to facilitate, support or promote outward FDI – in other words, to help firms establish foreign affiliates.

B The different categories of HCMs

The UNCTAD Secretariat conveniently defined six types of HCMs. First, there are 'policy positions'. These fairly common HCMs are defined by UNCTAD as 'positive in tone but vague in specific commitments [...] generalized statements on intentions or goals'.[91] These are mostly non-binding statements/declarations in which states are expressing their general intention to support outgoing investment. This type of HCM is fairly common under international law and can be found mostly in treaties' preambles.

The second type of HCM is 'information provision and technical assistance'.[92] Home states' collection and dissemination of information concerning host states' legal systems, economic environment and political situation is vital for prospective investors, and useful for the facilitation of foreign investment. As discussed below, international law instruments (e.g. Multilateral Investment Guarantee Agency (MIGA), investment treaties) are providing this type of HCM.

A third type of HCM is defined as 'technology transfer'.[93] These are measures that are aimed to support specific types of investments; those that result with the transfer of technologies to the developing world. This type of HCM can appear in different shapes, whether through direct support of technology-intensive investments (e.g. via grants) or by the facilitation of such investment through capacity building and technical assistance that will enhance host states' receptive capacity.[94] Examples of international law tools that are falling under this category are the Kyoto Protocol's Flexible Mechanisms, which are providing financial incentives for such investments,

90 Karl P. Sauvant et al. 'Trends in FDI, home country measures and competitive neutrality' in Andrea K. Bjorklund (ed.) *Yearbook on International Investment Law & Policy 2012–2013* (OUP 2014), 11–12.

91 UNCTAD (n 89) 8.

92 UNCTAD (n 89) 9.

93 UNCTAD (n 89) 9.

94 UNCTAD (n 89) 9.

and (debatably) also the TRIPS Agreement, which requires host states to adopt basic standards for IP protection.

The fourth type of HCM in UNCTAD's typology is 'financial and fiscal incentives'. This group of HCMs includes a variety of measures, *inter alia* grants, loans and a myriad of taxation measures.[95] This type of HCM is provided through a variety of international law instruments, ranging from the Kyoto Protocol's Flexible Mechanisms, to Official Development Assistance (ODA), subsidised state-owned insurance-based tools and rarely even investment treaties.

The fifth HCM in the UNCTAD's typology is defined as 'investment insurance', somewhat narrowly defined as the use of insurance mechanisms provided by agencies such as the World Bank's Multilateral Investment Guarantee Agency (MIGA) or domestic branches such as the US' Development Finance Corporation (DFC, formerly the Overseas Private Investment Corporation (OPIC)) or the UK Export Finance (UKEF). The UNCTAD refers mostly to 'political and other risks that most private insurance companies will not cover'[96] and adds '[s]uch risk is generally higher in developing countries, so the practical effect of these HCMs is to support FDI in developing countries'.[97] As already stated, the most striking example of this international law HCM is the MIGA Convention and the numerous bilateral insurance agreements concluded by the US with countries in which DFC/OPIC operates. A wider understanding of the term 'insurance' might imply that investment protection treaties could also be included under this title, as these instruments provide a certain 'insurance' against political risks, especially where equipped with an effective enforcement mechanism in the shape of investment arbitration.

The sixth category of HCM is 'market access regulations', including 'trade-related measures dealing with matters such as product certification, country-of-origin definitions or preferential import regimes'.[98] These measures facilitate foreign investment and reducing costs. Sauvant et al. are referring to this type of measures as 'indirect' measures, as they do not address the process of investment directly.[99] However, as elaborated below, such measures will be addressed as 'direct' HCM in this study, because in today's economy activities such as trade and investment are often entangled and inseparable. In international law, this type of HCM is found primarily in trade agreements.

The last category of HCM is 'extraterritorial controls', i.e. the subjecting of investors who are operating in foreign markets to their *home countries* regulations. It is questionable whether such measures should be considered

95 UNCTAD (n 89) 10.
96 UNCTAD (n 89) 37
97 Ibid.
98 UNCTAD (n 89) 10.
99 Sauvant et al. (n 90) 19.

as HCMs as they do not necessarily encourage Foreign Direct Investment (FDI). In fact, the UNCTAD's study on HCMs specifically states that the business community dislikes such measures as they may create conflicts with host states' jurisdiction.[100] Regardless, such measures can be found under different types of international law, the most striking example being soft law guidance that is incorporated under other HCMs (e.g. MIGA) and provides certain supervision over the activity of the investor. An even more striking example is the Zero Draft of a treaty on business and human rights (officially titled: 'legally binding instrument to regulate, in international human rights law, the activities of transnational corporations and other business enterprises'[101]), currently negotiated within the UN. If accepted, foreign investors' violation of human rights abroad will be subject to the jurisdiction of home countries' courts.

C Why HCMs?

As stated above, there are many answers to the 'how' question. This study, however, will focus on international law HCMs. Naturally, the decision to focus on HCMs implies neglect of an important part of international law, namely, measures that are taken by *host* states for the purpose of enticing foreign investors to invest in their markets. The decision to limit the scope of this study was taken for several reasons.

To begin with, the focus on HCMs implies an emphasis on *investment exporting countries*. In other words, the focus is mostly on wealthy nations, and the policies that these states could adopt in order to promote sustainable investment by their own nationals. Importantly, these nations – and their investors – have the means to invest in objectives such as the abatement of climate change, or the attainment of SDG 7. The financial/resources capability of these states is important under international law for several reasons. First, having the means implies also a power to make a difference. Targeting policies in investment exporting countries is effective in this respect; HCMs are designed to directly affect the behaviour of rich countries' investors, and have the potential of re-directing investment decisions towards desirable destinations. As explained in Chapter 1, much of the expected financing pledges in the context of climate change are expected to arrive from the private sector – notably from these very investors.

Moreover, as stated below, some of the recommended HCMs also involve a certain financial contribution by investors' home state (e.g. grants, guarantees, insurances). The wealth of investment exporting states implies

100 UNCTAD (n 89) 45.

101 Zero Draft 'Legally binding instrument to regulate, in international human rights law, the activities of transnational corporations and other business enterprises' (July 2018) <https://www.business-humanrights.org/sites/default/files/documents/DraftLBI.pdf>.

the ability to invest sufficient resources in these HCMs, and consequently, increases their effectiveness. This is all in line with the principle of common but differentiated responsibilities, which has, in recent years, become the centrepiece of the international efforts to address climate change and development.[102] This principle demands higher levels of responsibility and accountability from the wealthier nations. This augmented responsibility goes beyond the obligation to increase domestic efforts. It also implies an imperative to assist developing nations in their own efforts to combat climate change and, more broadly, achieve the SDGs. HCMs could potentially be used in order to fulfil this obligation because, as stated above, investment exporting states could use these measures in order to channel RE-related investment towards developing markets.

Third, the focus on HCMs represents a step forward in the discussion on policy integration. The focus of many of existing policy-debates (notably in areas such as investment and trade law) is on 'defensive' measures, ensuring host states' ability to protect local interests and securing them a sufficient regulatory flexibility. It is time, however, to take a step forward beyond this discussion, and address the far more complex world of *active* integration, i.e. the development of mechanisms that will promote 'integrated' investment, rather than simply curtailing non-sustainable investment. An OECD report commented on this issue with respect to the integration of climate-related objectives in development policies[103]:

> Development co-operation activities should not merely do no harm to effective action on climate change. The activities should make a positive contribution to the system-wide transformation that is needed to achieve low-emissions, climate-resilient pathways. This is the extent of action necessary to ensure development co-operation delivers on its mandate and supports the achievement of the SDGs. It is insufficient for development co-operation providers to focus on meeting no more than the minimum standard of doing no harm by avoiding actions that actively undermine the Paris Agreement. Not all development co-operation activities need to include active climate objectives, but it is nonetheless critical that providers' underlying assumptions, conditions and objectives support a systemic, holistic approach to achieving low-emissions, climate-resilient development.

102 The principle of common but differentiated responsibilities is now entrenched in almost every document, decision or agreement issued by the UNFCCC See for example the UNFCCC preamble, as well as Articles 3(1), and 4 of this Convention, as well as Art 10 of the Kyoto Protocol, paragraph 1 of the Copenhagen Accord, the preamble to the Paris Agreement, and Art 2(2), 4(3), 4(19) of this agreement.

103 OECD, *Aligning Development Co-operation and Climate Action: The Only Way Forward* (OECD 2019) 21.

Fourth, this study's focus on HCMs is desirable for academic reasons. In short, current literature is focussed almost exclusively on *host countries'* measures and their efforts to attract foreign direct investment. Very little attention was granted to home country measures, a topic that was defined by Sauvant *et al.* as 'unexplored'.[104] This study is therefore expected to fill a significant gap in literature, and contribute to the existing body of knowledge. This gap in the literature is reflected also in states' practice: as reviewed below, international law HCMs are under-developed. Expending academic literature on this area will, hopefully, lead to increased attention towards the utilisation of international law HCMs, and unlock the potential that is embedded in this type of measures.

Finally, the reader should rest assured that 'defensive measures' such as safeguards were not entirely ignored in this study. It is, in fact, impossible to discuss HCMs in complete isolation from safeguards regarding issues such as human rights and environmental protection. Home countries' ability to impact these issues and impose necessary safeguards is simply too great to ignore, and indeed some of this study's chapters are addressing these issues.

D Why international law?

International law is, primarily, the regulation of states' conducts and states' relations at an international level. At the heart of this study are two problems that are *international* in nature – climate change and development in the developing world. Both of these problems include a strong element of international accountability, i.e. the international community, *as a whole*, has accepted responsibility for these issues and is attempting to resolve them.

The spirit of 'global partnership' is indeed at the heart of global agenda and politics, at least in the context of development and climate change. Target 17.16 of SDG 17 instructs the international community to 'enhance the global partnership for sustainable development complemented by multi-stakeholder partnerships that mobilize and share knowledge, expertise, technologies and financial resources to support the achievement of sustainable development goals in all countries, particularly developing countries'.[105] The UNFCCC is equally '[a]cknowledging that change in the Earth's climate and its adverse effects are a common concern of humankind', as well as that 'the global nature of climate change calls for the widest

104 Karl Sauvant et al. 'Trends in FDI, home country measures and competitive neutrality' in Andrea Bjorklund (ed.) *Yearbook on International Investment Law & Policy 2012–2013* (OUP 2014) 3 <http://ccsi.columbia.edu/files/2014/03/Yb-12-13-ch.-1-8-Nov-13-stand-alone-final-for-website.pdf>.

105 See Target 17.16 of SDG 17, United Nations General Assembly, *Transforming our World: The 2030 Agenda for Sustainable Development* (2015) A/RES/70/1.

possible cooperation by all countries and their participation in an effective and appropriate international response'.[106]

More specifically, while the operationalisation of such 'global partnerships' is admittedly vague and debated, it is clear that this cooperation's blueprint, as prescribed by international law, is based on the assumption that resources and expertise from the developed world should be used in order to support the efforts of the developing world. This dynamic is not only rooted in international law; it is also enabled, and executed, by this legal system. It is international law that is expected to deliver on this mechanism of global accountability and development, and, indeed, it is primarily within this framework that efforts are being made to enable it.

The cross-border, North-South element discussed above is especially notable in the context of this study. Only through *international* cooperation will sufficient RE-related investment and technologies flow to the Global South. Moreover, in light of the significant investment barriers reviewed in Chapter 1 of this study, only through *public* intervention will such investment materialise. The focus on international law is therefore a necessity; it represents a *public* intervention that enables a *cross-border* activity, and aims to achieve an objective defined as 'a common concern of humankind' and as an international goal for humanity as a whole.

And what of national/domestic laws and policies? Such laws and policies are, obviously, extremely important. This study's sole focus, however, is on international law. As stated above, the legal universe that regulates investment in RE is enormous and addressing it in its entirety is far beyond the scope of this research. Others will, hopefully, complement this study with reviews of national and municipal laws.

The focus on international law HCMs, however, is justified by more than the author's mere time and scope restrictions. As explained below, domestic HCMs reflect the interests of the home country alone; they are designed exclusively by the home country, and are controlled by it. International law HCMs, on the other hand, represent a negotiated arrangement that reflect a common set of interests, are jointly designed and are controlled by both home and host countries. As such, they represent a more equitable solution, which represent a far wider range of public interests. This last point will be discussed further below.

E The role of international law HCMs

The above section explains the importance and relevance of international law, and HCMs, to this study's objective. The next task is to combine the two and define exactly what 'international law HCMs' are. The term HCMs was explained above, as 'policy measures taken by the home countries of

106 See preamble to the UNFCCC.

firms that choose to invest abroad designed to encourage FDI flows to other countries'.[107] Previous attempts to map the different types of HCMs imply that most of these measures are *unilateral* and *domestic* in nature, in the sense that they are offered domestically through national laws.[108] Sauvant *et al.* address in this respect mostly measures that are offered by governmental departments (e.g. ministries of foreign affairs, commerce and international trade) to local firms, including the provision of information, 'match-making', the provision of loans and guarantees, tax credit and investment insurance.[109] The design and the operationalisation of HCMs in this respect are regarded as an entirely national matter, aimed to promote local industries' interests and the local economy more widely.

International law HCMs, on the other hand, 'are those that were created in a treaty between two (or more) states, following negotiations, and are based on the notion of international cooperation'.[110] International law HCMs are different from 'unilateral' HCMs in several ways. First, the objectives of international HCMs are fundamentally different from those of 'unilateral' HCMs and are far wider in nature. Elsewhere, the author of this study elaborated on this issue[111]:

> Notably, unilateral HCMs incorporate mostly economic and commercial objectives; they are meant to support local industry in its efforts to increase its wealth through international expansion, and indirectly also the well-being of the supporting state. International law HCMs on the other hand, are intended to support much wider and more public objectives; notably, the well-being of *all involved* states, and in some cases even more. As such, international law HCMs often include objectives that are related to sustainable development, poverty alleviation, technology transfer, and as discussed in this paper, also to environmental protection. Increasingly, these HCMs are also designed with careful attention to the host states' right to regulate, which is often regarded as restricted due to the involvement of international law.

One example that will be discussed in this study, in which the differences between the goals of HCMs and international law HCMs is striking, is the operation of public political risk insurers. Political risk insurances are offered *inter alia* unilaterally/domestically by governments through their dedicated national agencies. The alternative – international law HCMs – includes notably the MIGA Convention. MIGA's wider policy objectives are evident throughout this institution's operation. It supports only investments that

107 UNCTAD (n 89) 66.
108 See a review of HCMs in Sauvant et al. (n 104).
109 See table in Sauvant et al. (n 104) 14–15.
110 Kent (n 62).
111 Kent (n 62).

are 'developmentally sound and meet high social and environmental standards'.[112] According to the MIGA Convention's preamble, one of this Convention's objectives is 'to enhance the flow to developing countries of capital and technology for productive purposes under conditions consistent with their development needs, policies and objectives [...]'.[113] It is further stated in the official commentary on the Convention that MIGA 'should satisfy itself that the investment concerned will contribute to the economic and social development of the host country, comply with the laws and regulations of that country, and be consistent with the country's declared development objectives'.[114] The emphasis on the *host* state's well-being and development could not be clearer.

Another feature of international law HCMs is related to their design, which is far more complex than that of unilateral HCMs. This is mostly due to the wider range of objectives, interests and actors that must be considered. Unilateral HCMs tend to regard primarily one set of interests; the success of investors' activity abroad and the impact of such investment on home countries' economies. The tools that are used by governments' unilateral HCMs are equally straightforward – the de-risking of investments and supporting their commercial viability. The focus is almost exclusively on the investor and, for understandable reasons; states represent primarily the interests of their own nationals. Unless international law instructs otherwise, states will hardly ever adopt altruistic considerations vis-à-vis the interests of other countries' nationals. Even where states are unilaterally committing themselves to address host states' interests (e.g. through environmental and social codes), these commitments are often soft in essence, and, naturally, are not the result of discussions with host states' concerning their genuine needs and interests.

The design of international law HCMs is an entirely different matter. These measures are based on international cooperation and negotiations, and therefore require the consideration of a much wider set of interests. Importantly, international law HCMs are negotiated and agreed on by *all* states involved; as such they are aimed to reflect a much wider set of interests, *inter alia* host states' environmental protection, human rights, economic prosperity, and the investors' own financial interests. The fact that international law HCMs are the result of a joint effort implies another important advantage – both home and host countries have some level of 'ownership' over these mechanisms; they are expected to show more commitment to their success, and are more likely to support investment that is based on these HCMs.

Lastly, the development of international law HCMs reflects the increasingly 'distributive' and 'de-colonialised' direction that international law and

112 See: MIGA <https://www.miga.org/who-we-are>.

113 MIGA Convention (n 153), Preamble.

114 MIGA, *Commentary on the Convention Establishing the Multilateral Investment Guarantee Agency* (MIGA 1985) 11.

international relations are taking in areas such as climate change and development laws and policies. Developing countries are increasing their impact on forums of international governance such as the UNFCCC, the World Bank and the WTO. The days in which they were expected to sit quietly and receive the West's donations and intervention are long gone. Accordingly, where investment exporting nations are wishing to direct and support their home investors' activities in other countries, they can no longer expect to do so without any respect to these countries' voices and interests.

To conclude this argument, there are compelling reasons to explore the development of international law HCMs; examine their effectiveness and consider their improvement. Importantly, this study does not necessarily claim that states should abandon the use of unilateral HCMs. Realistically, it is hard to see states entirely letting go of the exclusive control of these measures and their ability to direct and apply control over their outward investment. On many occasions, the operation of unilateral HCMs is also the only available alternative. The development of international law HCMs should not be regarded as mutually exclusive to unilateral HCMs, but rather as tools that complete and improve states' efforts, as well as their legitimacy.

F What kind of international law HCMs?

The more specific evaluation of international law HCMs will be delivered in the rest of this book's chapters. The following section is not intended to replace this more in-depth examination. It is rather aimed to briefly present some of the options, and explain the more 'provocative' choices made in this study.

i Numerous alternatives

As reviewed in this study, there is a wide pool of examples/models based on which international law HCMs could be designed. This study will review several – notably those that are available via trade, investment, development, insurance and environmental treaties. Not all of these HCMs are available for investors. Some of them are gradually expiring, like the Kyoto Protocol, which provided economic incentives for investors in the RE sector and will be replaced – at some point – with a new mechanism that will be based on Article 6 of the Paris Agreement. Other examples are still lingering on the negotiations table for many years now – the WTO's Environmental Goods Agreement is one such example. More examples are available only to a limited number of states (e.g. those who included them in bilateral agreements), such as investment promotion measures that are set in bilateral investment agreement. The purpose of this study is not to establish the availability of international law HCMs, but more to point at available models that treaty-makers will be able to learn and draw inspiration from.

It is important at this stage to stress that the review conducted in this study is not exhaustive in nature, and more HCMs could be designed in order to support the activity of investors abroad. For example, taxation treaties could be designed in order to ensure favourable (or at least fair) taxation of investors in the RE sector. These treaties could also include a list of tax benefits (or additional tax for fossil fuel investors) that will reflect the real cost of carbon emissions and the societal benefits that are inherent in their mitigation.

Competition law is another area in which international law HCMs could be used effectively. For example, states could negotiate certain standards of competition law regulation in order to facilitate the activity of foreign investors abroad, not unlike the TRIPS Agreement which 'imposed' high standards of intellectual property on host states. Competition arrangements could be especially useful for RE investors, notably where state-controlled enterprises are involved.[115]

Other relevant ideas are mentioned in the OECD Better Policies for Sustainable Development report (2016).[116] Many of these are more likely to be implemented through unilateral measures but some – including with respect to reporting, public procurement, the greening of sovereign funds and more – could be addressed as part of international agreements as well.

It is clear that not all available international law HCMs should be used, at least not in their current form. Some international law HCMs are expected to be met with resistance, notably as they imply trade-offs that not all will be content with. Two such examples are the use of international economic law, and the reliance on Official Development Assistance. These objections deserve a brief discussion, which will hopefully clarify this study's choice to include them in its recommendations. This discussion is presented next.

ii The 'untouchable': international economic treaties

This study advocates the integration of international economic law instruments for the promotion of RE-related investment in the developing world. Not all, however, will agree with the deployment of some of the tools that this study recommends. Notably, instruments such as trade and investment treaties are seen by some as Western/corporate biased instruments that are aimed to serve wealthy, Western commercial interests through the exploitation of developing countries' resources. For example, as quoted elsewhere in this study, the UN Special Rapporteur on Human Rights and the Environment – Professor David Boyd – has commented in his 2018 report on the use of investment arbitration[117]:

115 OECD, *Better Policies for Sustainable Development* (n 83) 200.

116 OECD, *Better Policies for Sustainable Development* (n 83).

117 UNGA, *Report of the Special Rapporteur on the Issue of Human Rights Obligations Relating to the Enjoyment of a Safe, Clean, Healthy and Sustainable Environment* (2019) un Doc A/74/161 <https://undocs.org/en/A/74/161>.

> Corporations have used Investor-State dispute settlement mechanisms in investment treaties to file lawsuits seeking compensation for lost profits when climate policies are strengthened, creating regulatory chill. States should withdraw consent to arbitration or negotiate a carve-out for climate actions as a safeguard to protect against these types of lawsuits.

Extreme arguments in similar vein were also made against international trade treaties (notably the WTO's) and global financial institutions, essentially rejecting them as colonial powers reflecting the limited interests of the few, and calling for their destruction.

The answer for these critiques is complex. To begin with, the argument that economic globalisation was bad for the poor is (at least) lacking in nuance. The economic growth in Asians countries since the 1990s is telling in this respect: according to the World Bank's data,[118] 60% of the population in the East Asia/Pacific region lived in extreme poverty during the 1990s, making it the poorest region in the world. By 2013, this number dropped to only 3.5% of the population in this region. Together with the South-Asia region, the number of the extremely poor in these regions alone was reduced by more than 1 billion persons. Many of these developments are based on trade liberalisation, and driven by the rise of international economic law, notably the establishment of the WTO (1994).

Naturally, others (notably in Africa) did not enjoy similar success and it is hardly contested that the benefits of globalisation were not equally shared. But the claim that economic globalisation (and economic law) does not work for the poor, as explained, does not tell the entire story – it did (and does) work, and for many. While there is no doubt that globalisation resulted in both 'winners' and 'losers', one must not overlook useful tools that have worked in the past, just because of the fact that a utopic reality was not achieved. Rather, one must learn from the lessons of the past and see how successful tools could be refined and improved, in order to promote the objectives and the interests of the weakest among us.

Indeed, to a certain extent, it seems that African nations have chosen to take this route. In 2018, 44 African nations concluded an all-African trade treaty – the Agreement Establishing the African Continental Free Trade Area – which is expected to dramatically liberalise trade in goods and services in Africa. Regional economic integration is also happening for a while now in Africa,[119] and while aptitude for investment treaties is currently low,

118 World Bank 'No Poverty' http://datatopics.worldbank.org/sdgatlas/archive/2017/SDG-01-no-poverty.html; World Bank. *Atlas of Sustainable Development Goals 2018: From World Development Indicators. World Bank Atlas* (World Bank 2018) 2.

119 See review in Trudi Hartzenberg 'Regional integration in Africa' (2011) WTO Staff Working Paper No ERSD-2011-13 <https://doi.org/10.30875/fad9df15-en>.

some advancements are made also on this front.[120] A landmark achievement was made in this area – the 2016 Nigeria – Morocco investment treaty – hailed by some as 'one of the most innovative and balanced bilateral investment treaty (BIT) ever concluded'.[121]

Instead of addressing international economic law as an 'untouchable' instrument, policy-makers will do well to learn from the lessons of the past and consider relevant improvements that will channel the force of these treaties for the realisation of common goods.[122]

iii Reliance on official development assistance (ODA)

Many of the risks discussed in Chapter 1 of this study could be alleviated (or even eliminated) through the injection of public funds, including Official Development Assistance (ODA). For example, public funds could be used for the de-risking of investments (e.g. the subsidisation of insurance/guarantees), making them more affordable and attractive for private actors. Public funds could also be utilised to support initial upfront investments, and therefore reduce the impact of high-cost, long-term finance. Public funds could even be used for topping up returns, where high returns are expected in light of investment risks.[123] In short, states' strategic use of ODA funds could support foreign investment in the RE sector in many different ways.

The use of ODA as a financing tool for supporting private Western investors is understandably controversial. The rather visual contradiction that is created where ODA funds are spent in extremely poor areas, on projects that are directly aimed at the needs of rich Western cooperation, is naturally disturbing. One must, however, remember that there is a bigger perspective to consider, even if the result seems at first as cynical and distorted.

Paul Collier explains this wider perspective and the role that ODA funds could play in it.[124] The production of jobs is key in the process of fragile states' development; it creates a positive impact on societies' stabilisation and standards of living. The most important creator of jobs is the private

120 African nations are signed on more than 850 BITs, see review in UNECA, *Investment Policies and Bilateral Investment Treaties in Africa* (UNECA 2016).

121 Tarcisio Gazzini 'The 2016 Morocco-Nigeria BIT: An important contribution to the reform of investment treaties' *IISD Investment Treaty News* (26 September 2017) <https://www.iisd.org/itn/2017/09/26/the-2016-morocco-nigeria-bit-an-important-contribution-to-the-reform-of-investment-treaties-tarcisio-gazzini/>.

122 See for example Armand de Mestral and Celine Levesque, *Improving International Investment Agreements* (Routledge 2013).

123 See more options for the use of public finance in UNDP, *De-risking Renewable Energy Investment* (UNDP 2013) 35.

124 Paul Collier 'Fixing fragility: The role of the private sector and local institutions' in Brahima Coulibaly (ed.) *Foresight Africa: Top Priorities for the Continent in 2019* (Brookings 2018) 61.

sector, and especially foreign firms.[125] Attracting targeted foreign firms that are necessary for states' specific needs requires resources, and so, Collier explains, '[t]he future of development assistance is to use public money to offset the costs and risks of firms that pioneer new activities in poor and fragile societies'.[126]

The use of ODA for the promotion of green investment was also recommended in several OECD studies,[127] including in its above-discussed Better Policies report.[128] There are examples where ODA was successfully invested by states for the promotion of RE-related investment, leading to a line of direct and indirect benefits for host states' development.[129]

The author does not claim that the use of ODA for such purposes is free from problems. The point made here is that diverting ODA for the purpose of promoting investment in RE implies certain trade-offs – trade-offs that are in fact embedded in almost any climate policy. If the attainment of SDG 7 (access to energy) and SDG 13 (climate action) are to be regarded as a priority, these trade-offs will have to be accepted.

V Conclusion

This chapter commenced answering the 'how' question – how can the international community achieve the grand objective of increasing RE investment in the developing world? It reviewed two key terms/principles that will guide its answer to this question in the next five chapters that follow: *integration*, and *international law Home Country Measures.*

The choice to concentrate on these two terms was explained, and so was their content. Importantly, these terms reflect two realisations that are somewhat interlinked. The first is that trade-offs will be necessary. Trade-offs are inherent to the act of integration, and policy-makers will have to: first accept this reality, and, second, consider measures that will mitigate their adverse effects. The latter implies the imposition of a variety of safeguards that will address the impact that foreign investment has on host communities. These safeguards cannot be ignored: they are part and parcel of the process of foreign investment and, as such, Home Country Measures cannot be designed without them. The chapters that will follow address important safeguards, such as with respect to human rights and corporate social responsibility.

The second realisation that emerges concerns the responsibility of the industrialised world. The focus on HCMs is justified *inter alia* by the

125 As Collier explains, local firms 'typically remain informal to hide from government and consequently cannot reap economies of scale and specialization', Paul Collier (n 124) 61.
126 Paul Collier (n 124) 41.
127 OECD, *Development Co-operation Report 2012* (n 87) Chapter 13.
128 OECD, *Better Policies for Sustainable Development* (n 83) 205.
129 See for example Rogerio Studart and Kevin Gallagher 'Guaranteeing sustainable infrastructure' (2018) 255 *International Economics* 84, 89.

responsibility that is imposed on the Global North to actively support developing countries' efforts to address issues such as energy poverty and climate change. This responsibility (which links to the point on safeguards, made above) implies that developed countries must not only do more, but also that they can do so by supporting foreign investment and, ideally, their own economic interests. Improving international law HCMs in this respect is not necessarily an altruistic move – it is a measure that will improve the sustainability of their foreign investment, and therefore also its durability and long-term success. Seen from this perspective, it is clear that the challenge ahead is global in nature, and its success will be the success of the international community as a whole.

The following five chapters will take this discussion further by focussing on specific areas of international law. Each one of them will review international law HCMs that are available via any given field, and will assess the level of integration in which these measures have been designed.

3 International investment law

Looking beyond the controversies

I Introduction

The following chapters will discuss a line of international law Home Country Measures (international law HCMs). The first international law HCM to be discussed is found in the world of international investment law. International investment agreements (IIAs, or Bilateral Investment Treaties (BITs)) can support FDI in the RE sector in many different ways. The terms of IIAs can be roughly divided into two groups: investment protection and investment promotion. This chapter will address both types as international law HCMs.

Investment protection provisions could be regarded as falling under the UNCTAD's fifth category of HCMs – 'investment insurance'. These provisions provide investors with certain assurances against potential abuses of host states' power. Admittedly, most will not define investment protection measures as HCMs, but rather as tools that are designed to *attract* investment. Practice, however, reveals that in most cases it is the investment exporting countries that insist on invasive and strong investment protection provisions, with the purpose of protecting their own investors' activities abroad. Host countries, on the other hand, will often attempt to reduce the level of protection and even regard investment protection as encroaching on their sovereignty. This study therefore, defines investment protection assurances as international law HCMs.

Much has been written on investment protection provisions, including in the context of renewable energies (RE).[1] Nevertheless, the recent barrage of investment arbitrations addressing RE support schemes justifies a fresh review of the jurisprudence currently emerging on this matter. These decisions are crucial for any assessment regarding the effectiveness of investment protection provisions as HCMs. The first part of this chapter will therefore evaluate this emerging line of jurisprudence.

1 For example, Anatole Boute 'Combating climate change through investment arbitration' (2012) 35 *Fordham International Law Journal* 613; Edna Sussman 'The energy charter treaty's investor protection provisions: Potential to foster solutions to global warming and promote sustainable development' in Marie-Claire Cordonier Segger et al. (eds.) *Sustainable Development in World Investment Law* (Kluwer 2010) 513.

DOI: 10.4324/9780429277207-4

The second part of this chapter will assess the far less common (and far less explored) investment promotion provisions. As reviewed below, investment promotion provisions vary in their nature and shape and could be labelled under different categories of HCMs. The author believes that the potential of these provisions is great, and their utilisation should be increased.

II Investment protection under IIAs

Arguably the 'hottest' topic in investment law litigation in recent years has been disputes concerning investments in the renewable energy sector. While exact numbers are difficult to attain,[2] the International Institute for Sustainable Development (IISD) estimates that 40 RE-related arbitrations were launched against Spain alone.[3] Numerous other cases were submitted against other countries as well, notably against the Czech Republic and Italy. This large number of ongoing cases, and the jurisprudence that they will create, will determine the role of investment law as a tool for promoting future private investment in the RE sector.

A Investment in renewables: the need for assurances

RE investments are usually associated with a high upfront cost and a very long payback period.[4] For this reason, support schemes such as feed-in tariffs (FIT) are often designed to last between 15 and 25 years – a period sufficiently long to ensure the profitability of these investments. The problem with most support schemes, however, is that the same entity that makes the long-term promise to pay for the electricity – the state – is also responsible for enforcing the deal. In other words, once making the investment, the investor is entirely in the hands of the host state. As seen in a number of cases (discussed below), economic strains have led governments to re-open what had been considered, at least in the eyes of the foreign investor a 'closed deal'. The terms that investors expected (whether legitimately or not) were not respected, which leads to a drop in investment[5] and a negative impact on investors' confidence.[6]

2 Investment disputes are now always reported, or made known to the public.

3 Isabella Reynoso 'Spain's renewable energy saga: Lessons for international investment law ad sustainable development' (2019)10(2) *Investment Treaty News IISD.*

4 Lars Strupeit et al. 'Overcoming barriers to renewable energy diffusion: business models for customer-sited solar photovoltaics in Japan, Germany and the United States' (2016) 123(1) *Journal of Cleaner Production* 124.

5 Arjun Mahalingam and David Reiner 'Energy subsidies at times of economic crisis: A comparative study and scenario analysis of Italy and Spain'(2016) EPRG Working Paper 1603 <https://www.eprg.group.cam.ac.uk/wp-content/uploads/2016/02/1603-PDF.pdf>.

6 The Spanish Wind Energy Association claimed at the time that 'It is necessary to restore confidence before any new investments in wind energy are made in Spain' <https://www.aeeolica.org/en/press-room/aee-news/archive/3466-it-is-necessary-to-restore-confidence-before-any-new-investments-in-wind-energy-are-made-in-spain-according-to-the-spanish-wind-energy-association>; see also the European Court of Auditors

B Investment protection provisions as assurances for RE investors?

IIAs provide foreign investors with treaty-based guarantees against host states' malign actions. These treaties provide assurances against, *inter alia*, nationality-based discrimination, the abuse of regulatory powers, expropriation and more. The Investor-State Dispute Settlement (ISDS) system allows investors to enforce these guarantees independently and away from their host states' control through arbitral proceedings. These assurances are especially useful in the RE sector, where investments are typically long-term in nature; involve very high, upfront, sunk cost[7]; and rely almost exclusively on the states' will to uphold their side of the bargain.

It is clear that at least *potentially*, IIAs can influence investors' decision-making processes in favour of making an investment in the RE sector. *In theory*, the protections provided by IIAs should be most welcome by foreign investors; these treaties de-risk their investments, and ensure that acceptable standards of due process and rule of law are upheld.[8] Whether these guarantees are indeed influencing investors' decisions and leading to positive impacts on the promotion of Foreign Direct Investment (FDI), is, however, subject to a heated debate: it is questioned whether IIAs indeed increase foreign investment, or, in other words, whether the assurances provided by IIAs matter at all.[9]

The author of this study is not qualified to judge between the sides to this debate. A cursory review of the literature nevertheless reveals that most experts agree that IIAs do have a positive impact on FDI flows, even if only to a certain extent and with some limitations.[10] Interestingly, Collen

---

Special Report: Wind and Solar Power for Electricity Generation: Significant Action Needed if EU Targets to be Met (2019) <https://www.eca.europa.eu/Lists/ECADocuments/SR19_08/SR_PHOTOVOLTAIC_EN.pdf> para 50, 85.

7 See: Sujata Gupta et al. (eds) 'Chapter 16: Cross-cutting investment and finance issues' in *Climate Change 2014: Mitigation of Climate Change* (IPCC 2014) 1227.

8 Indeed several authors have argued that IIAs' assurances are supportive of RE-related investment. See, for example: Anatole Boute 'Combating climate change through investment arbitration' (2012) 35 *Fordham International Law Journal* 613; Edna Sussman 'The energy charter treaty's investor protection provisions: Potential to foster solutions to global warming and promote sustainable development' (2007) 14(2) *ILSA Journal of International and Comparative Law* 391. More recently the OECD identified the 'insufficient investor protection' as one of the reasons inhibiting a growth in low-carbon investment, see: OECD, *Better Policies for Development 2015: Policy Coherence and Green Growth* (OECD 2015) 73.

9 See review of the critical voices in Kyla Tienhaara 'Does the green economy need investment state dispute settlement?' in Kate Miles (ed.) *Research Handbook on Environment and Investment Law* (EE 2019) 292.

10 Aisbett et al. claim that IIAs promote FDI as long as claims were not brought against host states; Frenkel and Walter claim that strong ISDS provisions positively impact FDI flow (although Berger et al. argue that this impact is 'minor'); Berger et al. argue that where IIAs include liberal admission rules (e.g. National Treatment) they promote FDI and that 'RTAs without strong investment provisions may even discourage FDI'; Neumayer & Spess show a positive impact of BITs on FDI flows to developing countries, a conclusion

et al. found that the positive impact of IIAs is also dependent on the sector in which the investment was made, stating that 'FDI in those sectors with higher sunk costs responds more strongly to the signing of BITs'.[11] Investments in the RE sector fit this description, even more so than the average energy project. In fact, the upfront initial investment needed for a RE-based power plant is far higher than the equivalent fossil fuel-based alternative.[12] This is all suggesting that *prior to making an investment*, IIAs indeed play a positive role, especially in the context of investment in the RE sector.

Another indication in this respect is the effect of the emerging jurisprudence on RE-investment disputes. In the past, it was doubtful whether investors were even aware of IIAs' guarantees; whether these assurances have played any role in alleviating their concerns, and ultimately affecting their decision to invest. At the time of writing, however, there is little doubt that RE investors are conscious of IIAs' guarantees, as well as the instrumental role that these assurances have had in recovering very significant amounts. After dozens of concluded cases,[13] and millions ordered in compensation, how could they not? The substantial rise in investment claims is therefore playing an important role in enforcing the role that IIAs are playing in RE investors' decision-making process, prior to making their investment.

C Enforcing assurances through ISDS

It seems then likely that even with all the necessary caveats, the assurances provided by IIAs are useful in the RE sector. As stated above, the ability to

---

that was re-confirmed by Busse et al., which add that they 'may even substitute for weak domestic institutions, though probably not for unilateral capital account liberalization'. See more in a literature review of studies that dealt with this question in Zbigniew Zimny et al. 'The role of international investment agreements in attracting foreign direct investment' (2009) UNCTAD Series on International Investment Policies for Development (UN 2009), online: UNCTAD, <http://unctad.org/en/Docs/diaeia20095_en.pdf>. See Emma Aisbett et al. 'Bilateral investment treaties as deterrents of host-country discretion: the impact of investor-state disputes on foreign direct investment in developing countries' (2018) 154(1) *Review of World Economics* 119; Michael Frankel and Benedikt Walter 'Do bilateral investment treaties attract foreign direct investment?' (2018) WHU Working Paper 17/08 <https://opus4.kobv.de/opus4-whu/frontdoor/index/index/docId/537>; Axel Berger 'Do trade and investment agreements lead to more FDI? Accounting for key provisions inside the black box' (2013) 10(2) *International Economic and Economics Policy* 247; Eric Neumayer and Laura Spess 'Do bilateral investment treaties increase foreign direct investment to developing countries?' (2005) 33(10) *World Development* 1567; Matthias Busse et al. 'FDI promotion through bilateral investment treaties: More than a bit?' (2010) 146(1) *Review of World Economics* 147.

11 Lisbeth Colen et al. 'Bilateral investment treaties and FDI: Does the sector matter?' (2016) 83 *World Development* 193.

12 Tobias Schmidt 'Low-carbon investment risks and de-risking' (2014) 4 *Nature Climate Change* 237.

13 See review in this chapter, below.

'redeem' these assurances is currently being tested by investors in this sector in numerous investor-state arbitrations. Nevertheless, does investment law really lives up to these expectations? The answer to this question will be known once the dust has settled and the dozens of investment awards in RE disputes have been published. For the time being, the cases discussed in this chapter provide a strong indication as to what the answer will be.

The following review will analyse the usefulness of two legal guarantees that are commonly provided through IIAs: the protection from expropriation, and the obligation to provide fair and equitable treatment. The case-law review provided below is necessary for understanding the scope of protection provided by IIAs, as well as its content.

The below-discussed cases concern the withdrawal/reduction of benefits allegedly promised to investors in the RE sector. Many of these cases concern Spain and the Czech Republic, and more specifically these states' revocation of alleged promises made to investors in the RE sector. Other states (notably Italy but others as well[14]) have faced similar claims. A quick review of the facts that led to the claims against Spain and the Czech Republic is useful to understand the context of RE-related investment disputes.

D Spain's contested measures

The facts (which are portrayed in greater detail in the awards) are as follows: in order to meet its obligations under international and EU environmental laws, Spain adopted Law 54/1997. The Law's main objective was to create a 'special regime', which was supposed to support RE generation and encourage investment in this area.

Law 54/1997 was later concretised and supplemented with a line of regulations, in which specific tariffs were set for two defined periods (for the first 25 years of operation, and from year 26 onwards). During those years, Spain adopted a set of additional plans and engaged in promotional activities, all with the purpose of increasing RE-related investment. For example, in 2005 the Spanish Government adopted its 2005–2010 RE Plan (PER 2005–2010) in which ambitious targets were set ('29.4% of the electricity generation from renewable resources'[15]). The plan included *inter alia* statements such as that 'the proper functioning of these mechanisms must be guaranteed [...] to maintain investor's confidence' and that 'the Special

14 See for example ongoing arbitrations against Argentina (Orazul International Espana Holdings SL vs Argentine Republic) and Romania (e.g. *Wind Project vs Romania*; *LSG Building Solutions GmbH and others vs Romania*), or concluded arbitrations against Albania (Albaniabeg Ambient Sh.p.k and M. Angelo Novelli, and Costruzioni S.r.l.v. vs Albania), Bulgaria (ENERGO-PRO v. Bulgaria,EVN v. Bulgaria; other cases still ongoing) and other states as well.

15 Novenergia award, para 104.

Regime should maintain investor's confidence [...] through a stable and predictable support scheme'.[16]

The Spanish attempt to attract investment in its RE sector was highly successful. But then the year 2008 arrived, and with it the global financial crisis.

Spain was one of the countries most drastically affected by the global financial crisis. The generous support schemes for the production of RE which were once considered as a priority, became, almost overnight, a dispensable luxury. The economic crisis led to a drop in consumers' electricity demand, and accordingly also to a drop in the funds available for financing the RE support schemes. These developments led to a substantial financial deficit in the authorities' budgets.

The regulatory approach changed accordingly. New laws were enacted, effectively remaking the terms and conditions that were promised to investors in the RE sector. Notably, new tariffs were introduced as well as new technical requirements, a new tax, and a cap on the number of eligible production hours.

Some investors attempted, unsuccessfully, to challenge this regulatory reform before domestic courts. Eventually however, many of these investors resorted to international investment law – the Energy Charter Treaty (ECT) more specifically – demanding compensation from the Spanish government for what, they claim, was essentially a breach of the promises that led them to invest in Spain in the first place.

Despite these challenges, Spain's modification of its RE regulation did not stop. By 2013 (RDL 2/2013, RDL 9/2013 and Law 24/2013), Spain effectively terminated the 'special regime's' Feed-In Tariffs (FIT) scheme, replacing it with a 'special payment' regime, according to which producers are entitled to a 'reasonable return'. While the FIT paid producers per unit of electricity, the 'reasonable return' is calculated based on the value of the initial investment, operation costs and revenues. The preamble to Law 24/2013 provides an indication as to the reasons for these regulatory changes[17]:

> Despite the fact that tolling increased by [one hundred and] twenty two percent between 2004 and 2012, positioning the electricity price in Spain well above the European Union average, this was not enough to cover the system's costs. This imbalance has reached the point where the accumulated debt of the electrical system is currently in excess of twenty six billion Euros, the structural deficit of the system stands at ten billion per annum and the failure to correct the imbalance has introduced the risk of the bankruptcy of the electrical system.

16 Novenergia award, para 105.

17 *RREEF Infrastructure (G.P.) Limited and RREEF Pan-European Infrastructure Two Lux S.à r.l. vs Kingdom of Spain* (Decision on Responsibility and On the Principles of Quantum, 30 November 2018) ICSID Case No. ARB/13/30, para 135.

> Law 54 enacted on November 27th 1997 has proven insufficient to ensure the financial balance of the system, amongst other reasons because the remuneration system for regulated activities has lacked the flexibility required for its adaptation to major changes in the electrical system or in the evolution of the economy.

E The Czech Republic's contested measures

A similar scenario – albeit with important different details – took place also in the Czech Republic, leading to a barrage of cases against this country.[18] Since the year 1992, the Czech Republic have been adopting a line of support schemes that were aimed at the promotion of investment in RE. The first key step was the adoption of two tax incentives for RE producers (1992). A second important step was the adoption of Act No. 180/2005 ('Act on Promotion', 2005), which aimed to 'promote the exploitation of renewable energy sources'. The Act on Promotion introduced a series of incentives, including: '(1) preferential treatment of RES producers in the distribution or transmission of electricity, (2) fixed purchase prices or Feed-in-Tariffs ("FIT") and, alternatively, (3) Green Bonuses ("Green Bonuses", and together with the FIT, "Subsidies" or "Tariffs")'.[19]

The Czech FIT was initially set for 15 years, with a purchase price fixed annually by the Czech authorities. The Act on Promotion also installed restrictions on the government's ability to change the tariffs,[20] and, importantly, tied the government's ability to reduce tariffs to no more than 5% per year, ensuring long-term price stability. Another guarantee that was made through the Act on Promotion was that the FIT's payments were set to enable investors to recover their investment within 15 years (a promise that was repeated also in a Technical Regulation that was issued later on), as well as a 7% annual profit during this term.[21]

Government officials have released a number of statements concerning the Act on Promotion with the objective of convincing investors of its usefulness:

> [i]n particular, Mr. Martin Bursík, one of the co-authors of the Act on Promotion, who became Minister of Environment from 2007 to 2009, stated in his article dated 1 June 2005 that the most important principle

18 See *inter alia Antaris vs The Czech Republic; Natland v. Czech Republic; Jürgen Wirtgen, Stefan Wirtgen, Gisela Wirtgen and JSW Solar (zwei) GmbH & Co. KG v. Czech Republic.; I.C.W. Europe Investments Limited vs The Czech Republic; WA Investments-Europa Nova Limited vs The Czech Republic; Voltaic Network GMBH vs The Czech Republic; Photovoltaik Knopf Betriebs-GMBH vs The Czech Republic.* All awards were issued on May 15 2019, by the same tribunal.

19 *Photovoltaik Knopf Betriebs-GMBH vs The Czech Republic* at Para 136.

20 *Photovoltaik Knopf Betriebs-GMBH vs The Czech Republic* at Para 140.

21 *Photovoltaik Knopf Betriebs-GMBH vs The Czech Republic* at para 446.

> of the Act on Promotion was 'the guarantee of a stable feed-in tariff for a 15 year period following the launch of the power station into operation'.[22]

In 2009 the Czech authorities adopted the Pricing Regulation, in which an annual increase of 2%–4% to the FIT was determined and guaranteed for a period of 20 years.

The 2008–2009 financial crisis brought political turmoil. The then Czech Prime Minister resigned and a new temporary government was appointed. During this period, and in light of the obvious economic strains, the authorities began publicly expressing the view that the FIT prices were too high. These comments were aimed especially at the solar market, where production prices were significantly reduced and, accordingly, profit margins soared.

In August 2009, the government announced its intention to abolish the 5% price reduction restriction. The massive increase in the solar sector's profitability (due to a reduction in production cost) was mentioned as a reason for the annulment of this change.[23] In March 2010, the 5% restriction was abolished, but only for certain investments: 'for those types of renewable resources, which, in the year in which the new feed-in tariffs are being determined, achieve the investment return shorter than 11 years'.[24]

In October 2010 further changes were made: the 1992 tax incentives were cancelled, and a new solar levy was imposed on 'facilities commissioned during the period from 1 January 2009 to 31 December 2010'.[25] In 2012, all existing FIT contracts were to be terminated, and new, more onerous contracts were imposed on RE firms. Existing FIT tariffs, however, were not affected.[26]

F An emerging jurisprudence?

As stated above, there are still many ongoing arbitrations and it could be risky to predict at this point the emergence of a clear line of jurisprudence. Nevertheless, it is equally hard to ignore the strong indications that are already evident, most of which are pointing at a very specific direction. While investment tribunals do not follow the *stare decisis* rule, these indications are useful for investors wishing to understand the value, and the scope, of the assurances provided by IIAs.

22 *Photovoltaik Knopf Betriebs-GMBH vs The Czech Republic* at para 141.

23 Czech government announcement, as quoted in *Photovoltaik Knopf Betriebs-GMBH vs The Czech Republic* at para 159.

24 *Photovoltaik Knopf Betriebs-GMBH vs The Czech Republic* at para 173.

25 *Photovoltaik Knopf Betriebs-GMBH vs The Czech Republic* at para 172.

26 *Photovoltaik Knopf Betriebs-GMBH vs The Czech Republic* at para 175.

Claimants made several legal arguments against countries such as Spain, Italy and the Czech Republic. This chapter, however, will discuss only those claims that were made based on two IIAs' assurances: expropriation, and the fair and equitable treatment (FET) principle. These two standards of protection were at the heart of these disputes, and these are the relevant IIAs' provisions that investors should pay most attention to. Importantly, these two standards of protection exist also in many other IIAs around the world. The lessons learned from these cases may therefore be of great importance elsewhere as well.

i Expropriation

Most IIAs today provide strong protection against uncompensated expropriation.[27] There are different types of expropriation. First, there is a 'direct' form of expropriation, defined as 'a mandatory legal transfer of the title to the property or its outright physical seizure'.[28] This type of expropriation is fairly uncommon today, albeit not unheard of in the context of energy-related investment.[29]

In the case of clear, direct expropriations, there is no doubt that IIAs will be of use for RE investors. The protection of foreign investors from uncompensated expropriation is straightforward, both under IIAs, and, arguably, under customary international law.[30] Cases of direct expropriation, however, are relatively infrequent and thus of less interest for the purpose of this study. Indeed most of the claims in RE-related disputes to date have focussed on the more common, yet complicated, 'indirect' form of expropriation.

There is no one acceptable definition of the term 'indirect expropriation'.[31] The *Tippets v. Iran* Tribunal famously defined indirect expropriation as:

27 According to an UNCTAD report, 'Today virtually all bilateral investment treaties (BITs) contain an expropriation provision. Customary international law also contains rules on the expropriation of foreign owned property and continues to supplement IIAs on those issues where the latter leave gaps or require interpretation'. UNCTAD, *Expropriation* (UNCTAD 2012) <https://unctad.org/en/Docs/unctaddiaeia2011d7_en.pdf> 5.

28 UNCTAD (n 27) 6.

29 Clifford Krauss 'ConocoPhillips wins $2 billion ruling over Venezuelan seizure' (25 April 2018) *New York Times* <https://www.nytimes.com/2018/04/25/business/energy-environment/conocophillips-venezuela-ruling.html>; BBC News 'Argentina to expropriate Repsol oil subsidiary YPF' (17 April 2012) *BBC News* <https://www.bbc.co.uk/news/business-17732910>.

30 Concerning the status of this legal rule under customary international law, see Stephen M Schwebel 'The influence of bilateral investment treaties on customary international law' (2004) 98 *American Society of International Law Proceedings* 27; for an opposing view, see Matthew C. Porterfield 'State practice and the (purported) obligation under customary international law to provide compensation for regulatory expropriations' (2011) 37 *North Carolina Journal of International Law and Commercial Regulation* 159.

31 OECD '"Indirect expropriation" and the "right to regulate" in international investment law' (2004) OECD Working Papers on International Investment 2004/04 <https://www.oecd.org/daf/inv/investment-policy/WP-2004_4.pdf> 5.

> [a] deprivation or taking of property may occur under international law through interference by a state in the use of that property or with the enjoyment of its benefits, even where legal title to the property is not affected.[32]

Sloane and Reisman further explain:

> [F]oreign investments may be expropriated 'indirectly through measures tantamount to expropriation or nationalization'. [FN omitted] This phrase, contained in the United States-Russia BIT and mirrored in substance in virtually all BITs, includes, not simply intentional and obvious indirect expropriations, nor only intentional creeping expropriations, a frequent form of taking in prior generations. It also captures the multiplicity of inappropriate regulatory acts, omissions, and other deleterious conduct that undermines the vital normative framework created and maintained by BITs – and by which governments can, in effect but not name, now be deemed to have expropriated a foreign national's investment.[33]

It is hardly surprising that investors have attempted, on several occasions, to claim that the revocation of benefits such as those 'promised' in FIT programmes have amounted to indirect expropriation. It is hardly questionable that the revocation of such benefits has indeed resulted in an 'interference by a state in the use of that property or with the enjoyment of its benefits'.[34] A more important question, however, is whether these measures are severe enough to be regarded as 'tantamount to expropriation or nationalization'?[35] This question was tested in a number of cases.

Many of the claimants in the RE-investment cases against Spain have argued that the financial impacts of the Spanish government's measures were so severe as to deem their investment 'expropriated'. For example, in *Infra-Red vs Spain* the claimants argued that 'the measures at issue drastically reduced the flow of income of the two (2) plants and resulted in a "substantial deprivation" of their investment – the standard to be applied in assessing the existence of an expropriation'.[36] In *9REN vs Spain* the claimants argued that 'Spain substantially interfered with 9REN's rights to receive the full value of the [FIT's benefits, AK]' and 'entirely deprived the Claimant of each of its rights under the aforementioned regimes'.[37] In *BayWa r.e. vs Spain* the claimants contended *inter alia* that they 'were no longer able to enjoy the

32 *Tippets v. Iran,* Iran-US Claims Tribunal 6 IRAN-U.S. C.T.R, 225.

33 W. Michael Reisman and Robert D. Sloane 'Indirect expropriation and its valuation in the BIT generation' (2004) 74 *The British Yearbook of International Law* 115, 118–119.

34 *Tippets v. Iran* (n 32) 225.

35 Reisman and Sloane (n 33) 118.

36 *InfraRed Environmental Infrastructure GP Limited et al. vs Kingdom of Spain* (Award, 2 August 2019) ICSID Case No. ARB/14/12, 481.

37 *9REN Holding vs Spain* (Award May 31, 2019) ICSID Case No. ARB/15/15, paras 353–355.

investment, as, since, 2011, the SPVs no longer pay dividends which "is in itself an expropriation"'.[38] More claims in this spirit were made in other cases as well.[39]

These claims, however, were overwhelmingly rejected. The *Novenergia* Tribunal held that the regulatory effect was not drastic enough to be considered as 'expropriation' and investors' property rights were 'left unaffected'.[40] The *Charanne* Tribunal stated that 'although the profitability of T-Solar could have been seriously affected',[41] the threshold for 'expropriation' requires a far more drastic impact than that created by the changes in the regulation:

> For a measure to be considered as equivalent to an expropriation, its effects must be of such a significance that it could be considered that the investor has been deprived, in whole or in part, of its investment. A simple decrease in the value of the shares constituting the investment cannot constitute an indirect expropriation, unless the loss of value is such that it can be considered equivalent to a deprivation of property.[42]

The *InfraRed* Tribunal relied on the *Electrabel v. Hungary* award in describing the high threshold that is required in cases of indirect expropriation:

> in both direct and indirect expropriation, consistently albeit in different terms, the requirement under international law for the investor [is] to establish the substantial, radical, severe, devastating or fundamental deprivation of its rights or the virtual annihilation, effective neutralisation or factual destruction of its investment, its value or enjoyment.[43]

The *InfraRed* Tribunal then stressed that the investors maintained control over their investment, and the investment remained profitable even after the changes in the regulation. There is therefore little doubt that the threshold for indirect expropriation was not met.[44] Other decisions have followed a similar line.[45]

38 *BayWa r.e. renewable energy GmbH and BayWa r.e. Asset Holding GmbH v. Kingdom of Spain* (Decision on Jurisdiction, Liability and Directions on Quantum, 2 December 2019), ICSID Case No. ARB/15/16, para 413.

39 See for example *Charrane vs Spain* (Award 21 January 2016) Arbitration No 062/2012 para 459; *Novenergia vs Spain* (Award 15 February 2018) SCC Arbitration (2015/063).

40 Novenergia Award (n 39) para 761. See also *Charanne* Award (n 39), para 462–467.

41 *Charrane vs Spain* (Award 21 January 2016) Arbitration No 062/2012 para 465.

42 *Charrane* Award (n 41) para 465.

43 *InfraRed* Award (n 36) para 505.

44 *InfraRed* Award (n 36) paras 505–509.

45 See for example *Isolux Infrastructure Netherlands vs Kingdom of Spain* (Award) SCC V2013/153; *Greentech Energy Systems A/S and others v. Kingdom of Spain* (Award, 14 November 2018) SCC Case No. V2015/150, paras 430–431; *Hydro Energy 1 S.à r.l. and*

These decisions are not surprising, nor are they controversial. Expropriation, by definition, is a very drastic measure and while investors often attempt to portray almost *any adverse* impact on their investment as 'expropriatory' in nature, tribunals will usually accept such claims only when the result is destruction/effective deprivation of the investment. This was not the case in these disputes and the claimants' expropriation arguments were accordingly rejected.

It can be learned therefore that even retrospective and meaningful changes to FIT programmes' terms and conditions, dramatic as these may be (and the Spanish example is possibly the most dramatic of them all), are highly unlikely to be seen as carrying a sufficiently drastic impact on RE investments so as to be legally regarded as 'expropriation'.

ii Fair and equitable treatment

While 'expropriation' claims are likely to fail, other claims – based on the requirement to provide Fair and Equitable Treatment (FET, (Art 10(1) ECT)) – have met with some level of success.

There is no one agreed-upon definition to the term 'fair and equitable treatment'.[46] Most, however, will accept that the FET secures investors' legitimate expectations at the time of making the investment, as well as their right to 'due process, transparency, freedom from coercion and harassment, stability, predictability and a general duty of due diligence'.[47]

The FET provision is possibly the most controversial and debated element in investment law. Enrique Boone Barrera described it elsewhere as too vague ('a riddle wrapped in an enigma') and called for its removal from investment treaties.[48] The FET's vagueness however, is not necessarily without reason or utility. It provides arbitrators with the required flexibility for addressing the endless and often creative ways, in which host states can mistreat foreign investors.

Investors involved in RE-related disputes have stressed the importance of stability in capital-intensive RE investment. Mostly, they based their arguments on the nature of the 'deal' that is inherent in the operation of support schemes such as FIT: the risky, upfront, substantial investment will be made only if sufficient financial guarantees are provided, especially with respect

Hydroxana Sweden AB v. Kingdom of Spain (Decision on Jurisdiction, Liability and Directions on Quantum, 9 March 2020), ICSID Case No. ARB/15/42, para 531–532.

46 Nicolas Angelet 'Fair and equitable treatment' Max Planck Encyclopaedias of International Law <https://opil.ouplaw.com/view/10.1093/law:epil/9780199231690/law-9780199231690-e2055?rskey=G5LHfN&result=1&prd=MPIL>.

47 Eric De Brabandere 'Fair and Equitable Treatment and (Full) Protection and Security in African Investment Treaties between Generality and Contextual Specificity' (2017) 18 *Journal of World Investment and Trade* 533.

48 Enrique Boone Barrera 'NAFTA's chapter 11 at a crossroad' International Law@UEA.

to the returns. Without such guarantees (e.g. duration and rates), investors would not have made their investment in the first place. Investors have emphasised in this respect their 'legitimate expectations' that support schemes will not be retrospectively changed, and that governments' representations made to attract investment in this sector will be respected.

The question of 'legitimate expectations' has been discussed by tribunals at great length. Notably, tribunals have addressed two, somewhat connected issues. The first is whether programmes such as FIT are specific enough to form 'legitimate expectations' for investors. The second question concerns states' (sovereign) right to change their laws at will in order to protect their public's interests, and whether this right must prevail over investors' expectations. These two issues will be discussed in more detail below.

SUPPORT SCHEMES: SPECIFIC ENOUGH TO FORM 'LEGITIMATE EXPECTATIONS'?

As mentioned above, most investors have based their FET claims on the term 'legitimate expectations'. More specifically, they argued that they 'legitimately expected' that the benefits promised by states through support schemes (notably FIT programmes) will remain in place for the duration explicitly specified in existing regulations. These support schemes, they argued, were created for the purpose of enticing investors by promising them predictability and stability, as specified in the detailed terms, rates and conditions, and as set in relevant laws and regulations when their investment was made. The specific promised terms and conditions, it is argued, were the basis on which the decision to invest was made.

Respondent states on the other hand, retorted that support schemes' 'promises' were made through laws and regulations. With the lack of an explicit freezing clause, investors had no right to expect that laws and regulations will remain unchanged: regulatory changes are, after all, a legitimate sovereign act. The states' position was based on the emerging, dominant line in investment law jurisprudence, according to which 'legitimate expectations' cannot be created through general legislation, but only through specific promises made to 'a particular group of investors about a specific benefit'.[49]

49 Teerawat Wongkaew, *Protection of Legitimate Expectations in Investment Treaty Arbitration: A Theory of Detrimental Reliance* (CUP 2019) 95. See review of the jurisprudence in there. See example of Tribunals that relied on this more nuanced approach *inter alia*: *Mobil Exploration and Development Inc. Suc. Argentina and Mobil Argentina S.A. v. Argentine Republic* (Decision on Jurisdiction and Liability, 10 April 2013), ICSID Case No. ARB/04/16, para 955; *TECO Guatemala Holdings, LLC v. Republic of Guatemala* (Award, 19 December 2013) ICSID Case No. ARB/10/23 paras 617–619; *Glamis Gold, Ltd. v. The United States of America* (Final Award, 8 June 2009) UNCITRAL para 813; *Grand River Enterprises Six Nations, Ltd. and others v. United States of America*, (Award, 12 January 2011) UNCITRAL para 141; *Mobil Investments Canada Inc. and Murphy Oil Corporation*

The important question to settle, therefore, is whether the benefits promised through RE support schemes – which are *specific* in nature, but delivered through *general* regulations – could be regarded as a source for investors' 'legitimate expectations'.

At the time of writing, the majority of tribunals have agreed that promises made through laws will usually be regarded as 'general' in nature, and thus subject to future changes. These tribunals nevertheless stressed that this is not a rigid rule, and that specific representations could be made, in some cases, also through the enactment of laws and regulations. The *Cube* Tribunal commented on the specific nature of FIT regulations:

> At least in the case of a highly-regulated industry, and provided that the representations are sufficiently clear and unequivocal, it is enough that a regulatory regime be established with the overt aim of attracting investments by holding out to potential investors the prospect that the investments will be subject to a set of specific regulatory principles that will, as a matter of deliberate policy, be maintained in force for a finite length of time. *Such regimes are plainly intended to create expectations upon which investors will rely*; and to the extent that those expectations are objectively reasonable, *they give rise to legitimate expectations when investments are in fact made in reliance upon them.*[50]

The *9REN* Tribunal has made similar observations:

> there is no reason in principle why such a commitment of the requisite clarity and specificity cannot be made in the regulation itself *where (as here) such a commitment is made for the purpose of inducing investment, which succeeded in attracting the Claimant's investment* and once made, resulted in losses to the Claimant.[51]

The *Antin* Tribunal equally commented:

> Given the precision and detail exhibited in the royal decrees, particularly the contemplation that the treatment would be accorded for a defined period of time, the Tribunal has no difficulty in concluding that

---

v. Government of Canada (Decision on Liability and on Principles of Quantum, 22 May 2012) ICSID Case No. ARB(AF)/07/4, paras 169–170. See also review of decisions in *Blusun S.A. vs Italian Republic* (Award 27 December 2016) ICSID Case No. ARB/14/3 paras 367–369.

50 Emphasis added. *Cube Infrastructure Fund SICAV and Others v. Kingdom of Spain*, (Decision on Jurisdiction, Liability and Partial Decision on Quantum, 19 February 2019) ICSID Case No. ARB/15/20, para 388.

51 Emphasis added. *9REN v. Spain* (n 37) 295.

> this falls squarely into the type of State conduct that was intended to, and did, give rise to legitimate expectations of the Claimants.[52]

A line similar to the cases discussed above, albeit with different outcomes, was taken in *Stadtwerke München vs Spain.* Here as well the Tribunal did not reject the notion that specific promises could be made through legislation.[53] However, based on its factual evaluation of the specific circumstances, the Tribunal decided (somewhat controversially) that a 'reasonable investor' would not have legitimately expected the FIT to remain unchanged.[54]

This line of decisions was followed also in other cases,[55] even if accompanied by a different tone. In *Blusun vs Italy* the Tribunal stated that: '[i]n the absence of a specific commitment, the state has no obligation to grant subsidies such as feed-in tariffs, or to maintain them unchanged once granted'.[56] On the face of it, this decision contradicts the line taken by tribunals in cases against Spain. A careful reading of the decision, however, reveals that the Tribunal's approach is not so different. Notably, the Tribunal stated that changes to law

> should be done in a manner which *is not disproportionate to the aim of the legislative amendment*, and *should have due regard to the reasonable reliance* interests of recipients who may have committed substantial resources on the basis of the earlier regime. *These considerations apply even more strongly when the context is subsidies or the payment of special benefits for particular economic sectors.*[57]

52 Emphasis added. *Antin Infrastructure Services Luxembourg S.À.R.L. and Antin Energia Termosolar B.V. vs Spain* (Award 15 June 2018) ICSID CASE No. ARB/13/31, para 552.

53 *Stadtwerke München GMBH, RWE INNOGY GMBH, and others vs Kingdom of Spain* (Award, 2 December 2019) ICSID Case No. ARB/15/1, 264.

54 *Stadtwerke München vs Spain* (n 53) 286. The Tribunal did not accept in this case even very specific presentations that were made by agencies such as InvestInSpain, which are (as described in the decision) 'wholly owned by the Secretariat of the State for Trade, and overseen by the Ministry of the Economic Affairs and Competitiveness'. The Tribunal's reasoning was that such agencies have no 'regulatory powers', and reasonable investors should not have relied on their representations. This explanation is controversial, as the Tribunal itself mentions that the Invest in Spain's official role is to attract foreign investment, and it was in this capacity that its representatives met with the investors. This will be tantamount to arguing that such agencies are acting independently, on their own will, and with no government oversight and instructions. Moreover, these agencies are created by the state with the purpose of attracting investors – surely the Spanish government itself would hope that prospective investors will take its own agency's promises at face value?

55 *Operafund Eco-Invest SICAV PLC and SCHWAB HOLDING AG vs Kingdom of Spain* (Award, 6 September 2019), ICSID Case No. ARB/15/36, para 485; *Novenergia vs Spain* (n 39).

56 *Blusun S.A. vs Italian Republic* (Award 27 December 2016) ICSID Case No. ARB/14/3 para 372.

57 *Blusun vs Italy* (n 56) 372.

The *Blusun* Tribunal, so it seems, confirms that more drastic changes to the law ('total alteration of the entire legal set-up for foreign investments', or changes that are 'completely dismantling the very legal framework constructed to attract investors'[58]) will not be tolerated, even where the original commitments were made through legislation. The distinction between this case and the cases above could therefore be understood in light of the different factual circumstances that existed in Italy, where the legal changes were not as dramatic as in Spain.

In *CEF Energia B.V. vs Italy*, the Tribunal reached the contrary result, ruling that Italy's regulatory change has breached the FET provision. While on the face of it this result is unique (most cases against Italy did not find a violation of the FET provision), a careful reading of the text suggests otherwise. Importantly, the circumstances were different in this case, and so was the Tribunal's justification. The Tribunal referred to tariff recognition letters that were sent to the claimant, in which it was specifically stated that incentives will last for a period of 20 years, with no explicit recognition of the state's right to unilaterally change this.[59] Moreover, the specific rates were listed in *specific contracts* concluded between the state and the investor.[60] The Tribunal considered the regulation that existed when the investment was made only as a secondary indication for investors' legitimate expectation, whereas the existence of specific agreements seems to be the decisive factor:

> No reading, no matter how indulgent, could lead anyone to consider anything other than a clear promise of twenty ears of constant currency incentives pursuant to a private law contract.[61]

The Tribunal explicitly distinguished this case from *Blusun* based on the fact that the source of expectations in this case was a private arrangement, rather than general legislation.[62]

The line according to which specific representation could, in principle, be made through general legislation was followed also in *Sunreserve vs Italy*.[63] In a decision that in effect followed the *Blusun*'s Tribunal's reasoning, the Tribunal agreed at first that specific promises could be made also via general

58 *Blusun vs Italy* (n 56) 363. The Tribunal is referring here to the obligation to provide stability, which is explicitly mentioned in the Energy Charter Treaty (ECT) but may not be explicitly mentioned elsewhere. However, the tribunal is relying in this section on case-law – which is not based on the ECT – and its determination seems valid to FET in a wider context.

59 *CEF Energia B.V. vs Italy* (Award, 19 January 2019) SCC Arbitration V (2015/158) para 212.

60 *CEF Energia B.V. vs Italy* (n 59) para 213.

61 *CEF Energia B.V. vs Italy* (n 59) para 217.

62 *CEF Energia B.V. vs Italy* (n 59) para 222.

63 *Sunreserve Luxco Holdings S.À.R.L. vs Italy* (Award 25 March 2020) SCC ARBITRATION V (2016/32).

legislation, only to reject the claimants' arguments based on a factual examination of the less-severe circumstances in Italy.[64]

A somewhat similar line is noticeable also in most of the decisions against the Czech Republic. In May 2019, four similar arbitral awards were issued by the same tribunal (the *WA/ICW/Photovoltaik/Voltaic vs The Czech Republic*).[65] In these decisions, the Tribunal did not reject the possibility that specific promises could be made via legislation. Nevertheless, following a careful analysis of the relevant facts, regulations and dates, it reached the conclusion that when the investment was made, regulatory changes were already known/reasonably predictable,[66] and therefore investors' reliance on existing regulation was not 'legitimate'.[67] Other tribunals agreed that some conditions (notably the duration of the FIT and the minimum annual return) were indeed specifically promised, but as these benefits were in fact provided, investors' expectations could not be regarded as frustrated.[68]

'PROMISES SHOULD BE KEPT' VS THE RIGHT TO REGULATE: FINDING THE RIGHT BALANCE

A second issue that played a significant role in RE-cases is the tension between, on the one hand, investors' expectations and demand for stability, and, on the other, the states' (sovereign) right to regulate and change their laws.

In essence, states have been arguing that their right to regulate should prevail over investors' expectations that FITs' benefits will not be changed. Spain, for example, has argued that 'stability does not mean freezing the legal regime and that even though the ECT limits the regulatory power of the signatory States, it neither cancels nor extremely curtails the State's regulatory power'.[69]

A similar claim has also been made by other states. The Czech Republic has emphasised in this respect 'the presumption that a host State does not violate its treaty obligation by simply changing its domestic law, noting that

64 *Sunreserve vs Italy* (n 63) 228–312.

65 *I.C.W. Europe Investments Limited vs The Czech Republic; WA Investments-Europa Nova Limited vs The Czech Republic; Voltaic Network GMBH vs The Czech Republic; Photovoltaik Knopf Betriebs-GMBH vs The Czech Republic.* All awards were issued on May 15 2019, by the same tribunal.

66 The Tribunal stressed that the investor was explicitly warned about these predicted changes, see *WA Investments-Europa Nova Limited vs The Czech Republic* (n 65) para 596–597.

67 *WA Investments-Europa Nova Limited vs The Czech Republic* (n 65) paras 543–563, 578–630.

68 *JSW Solar (zwei) GmbH & Co. KG vs The Czech Republic* (Award 11 October 2017) PCA Case No. 2014-03, paras 414–415.

69 *Antin vs Spain* (n 52) 527.

a State must be free to serve the public interest through legislative modifications'.[70] Investors, it is argued, are confusing

> the notion of stability with the distinct concept of stabilization, explaining that a guarantee of stabilization prohibits a host State entirely from changing its legislation, while a mere obligation to provide stability does not.[71]

Italy has taken a similar view, arguing that 'the legitimacy of investors' expectations must be assessed by taking into account and balancing the positions of, both, the foreign investor and the host State, by giving due relevance to the right of States to enact their legislations pursuant to their regulatory powers',[72] and that 'in the absence of a specific commitment to investors such as a stabilization clause, changes to regulatory framework established by means of general legislation do not qualify as a breach of legitimate expectations'.[73]

The 'Spanish' Tribunals have mostly accepted the investors' approach, at least to a certain extent. The *Eiser* Tribunal opened by stressing the importance of states' right to regulate, which includes also the right to adversely modify regulatory frameworks. In this case however, the Tribunal believed that Spain simply went too far. The change in the regulation was drastic, 'total and unreasonable'.[74] The Tribunal explains[75]:

> the evidence shows that Respondent eliminated a favourable regulatory regime previously extended to Claimants and other investors to encourage their investment in CSP. It was then replaced with an unprecedented [FN] and wholly different regulatory approach, based on wholly different premises. This new system was profoundly unfair and inequitable as applied to Claimants' existing investment, stripping Claimants of virtually all of the value of their investment.

The *RREEF* Tribunal agreed with the notion that the FET provision should not be understood as freezing states' right to regulate, and that it all depends on how drastic and radical states' changes are[76]:

> Stability is not an absolute concept; absent a clear stabilization clause, it does not equate with immutability. In this respect the Tribunal notes

70 *Photovoltaik Knopf Betriebs-GMBH vs The Czech Republic* at para 377.
71 *Photovoltaik Knopf Betriebs-GMBH vs The Czech Republic* at para 379.
72 *Sunreserve Luxco Holdings S.À.R.L. vs Italy* (Award, 25 March 2020) SCC ARBITRATION V (2016/32), para 628.
73 *Sunreserve Luxco Holdings S.À.R.L. vs Italy* (n 72) 632.
74 *Eiser* award para 364.
75 *Eiser* award para 365.
76 *RREEF Infrastructure vs Spain* (n 17) para 315.

> that the Claimants do not take such an extreme view. However, the obligation to create a stable environment certainly excludes any unpredictable radical transformation in the conditions of the investments.

In *Cube vs Spain*, the Tribunal followed a similar line, starting by stressing that investors could not 'reasonably rely on there being no change whatever to the Special Regime applicable to existing facilities'.[77] The Tribunal further agreed that investments made in 2011–12 cannot reasonably expect that the original tariffs (set by RD 661/2007) would not be modified, as discussions on their modification was already made public.[78] At the same time, the Tribunal stresses that '[s]uch changes, however, stop short of a radical alteration of the foundations of the regime that was then in place',[79] and that

> radical change in the nature of the Special Regime applicable to hydro plants already in operation under the Special Regime, amounting to an abandonment of the commitment to stability given in RD 661/2007, could not have been anticipated in 2010.[80]

In cases against Italy, a certain divergence is noticeable. As quoted above, the 'pro-state' *Blusun* tribunal placed the threshold for when the state's sovereignty shall prevail fairly high: at the 'total alteration of the entire legal set-up for foreign investments', or changes that are 'completely dismantling the very legal framework constructed to attract investors'.[81] A different Tribunal – *CEF Energia vs Italy* – has taken a more nuanced approach:

> The sustainability of the incentive system for PC producers and other valuable objectives of general interests were at stake, and it can readily be appreciated that the dignity which attaches so sovereigns must be given deference, though that is not infinite when balanced against a combination of international obligations freely assumed and, further, fact-specific circumstances of the extent of commitments given to investors.[82]

77 *Cube Infrastructure Fund SICAV and others v. Kingdom of Spain*, (Decision on Jurisdiction, Liability and Partial Decision on Quantum, 19 February 2019) ICSID Case No. ARB/15/20, Para 334.

78 *Cube* (n 77) para 350.

79 *Cube* (n 77) para 353.

80 *Cube* (n 77) para 354.

81 *Blusun vs Italy* (n 56) 363. The Tribunal is referring here to the obligation to provide stability, which is explicitly mentioned in the Energy Charter Treaty (ECT) but may not be explicitly mentioned elsewhere. However, the tribunal is relying in this section on case-law – which is not based on the ECT – and its determination seems valid to FET in a wider context.

82 *CEF Energia B.V. vs Italy* (Award, 19 January 2019) SCC Arbitration V (2015/158) para 240.

While most[83] cases against the Czech Republic tribunals have ruled against the investors on this point, it seems that their approach was nevertheless similar to that of the 'Spanish' tribunals mentioned above. The Tribunal in the *WA/ICW/Photovoltaik/Voltaic* cases opened by agreeing with the 'Spanish' tribunals that the Czech incentive regime cannot be seen as set in stone.[84] Indeed, modifications were introduced over the years to this scheme (e.g. the expansion of the plants' lifespan from 15 to 20 years). The Tribunal mentions especially the drastic change in circumstances (i.e. the significant reduction in production prices and the 2008–2009 economic crisis) that legitimately merited a regulatory change.[85] The Tribunal further indicated that at the time of making these investments the government had already made its intentions to amend the FIT's conditions publicly known[86]; a fact that suggests that the government acts were not unreasonable.

While this line of decisions seems fairly coherent, future investors should be aware that alternative views were expressed in the case-law. The above-discussed *Blusun* Tribunal's high threshold is one such example. Another notable example is Philippe Sands' dissenting opinion in *Operafund vs Spain.*[87] In essence, Sands considers the lack of an explicit stabilisation clause as a conclusive indication that the state's right to change its laws should prevail.[88] The fact that the law was indeed changed in the past (even if not as radically as the later changes) is a further indication that 'reasonable investors' should have considered.[89] Whether one agrees with Sands' view or not, future investors should be aware of this view and be mindful of the (for now, rather remote) possibility that the absence of an explicit stabilisation clause, even the most specific FIT regulation may not hold out against the state's right to change its laws.

In conclusion, investment tribunals, so it seems, are routinely balancing between the two conflicting interests. Importantly, while the right to regulate is repeatedly and decisively acknowledged, it is clear that most tribunals

83 At the time of writing there is one exceptional case, in which a Tribunal has ruled in favour of investors. In *Natland vs The Czech Republic* the Tribunal decided that the regulatory changes were contrary to investors' expectations and thus form a violation of the FET clause. At the time of writing this decision was not made public and available information is based on IAReporter's report of this decision, see Damien Charlotin 'Natland v. Czech Republic (Part 2 of 2): On the merits, tribunal finds stabilisation commitment in Czech legislation and breach of that commitment with introduction of solar levy' (26 July 2018) IA Reporter (access only with subscription).

84 See in *Photovoltaik Knopf Betriebs-GMBH vs The Czech Republic* at paras 449, 484.

85 *Photovoltaik Knopf Betriebs-GMBH vs The Czech Republic* at paras 512.

86 *Photovoltaik Knopf Betriebs-GMBH vs The Czech Republic* at para 475; *Voltaic Network GMBH vs The Czech Republic* at para 479; *ICW vs The Czech Republic* at Para 522; *WA Investments-Europa Nova Limited vs The Czech Republic* at para 563.

87 *Operafund Eco-Invest SICAV PLC and SCHWAB HOLDING AG vs Kingdom of Spain* (Dissent on liability and quantum, August 13, 2019).

88 *Operafund, dissent* (n 87) para 19.

89 *Operafund, dissent* (n 87) para 20.

do not consider this right as a 'carte blanche' for states to do as they will. This balancing act, so it seems, was performed on a case-by-case basis, and was influenced by certain factors. These factors are discussed next.

THE TYPE OF REFORMS, AND ITS IMPACT ON INVESTORS

The balance between states' right to regulate and investors' expectation was at the heart of most RE-related cases. The decisive factors in tribunals' decisions seem to be how radical the regulatory reform was, and how drastic its impact was on the investor.

As stated above, the *WA/ICW/Photovoltaik/Voltaic* Tribunal acknowledged that the regulatory environment indeed changed. The Tribunal, however, did not consider these changes as drastic. The Tribunal mentions that despite these changes, the reduction in solar panels' prices (and thus also in production costs) was so steep as to allow profits far and beyond those originally projected by the investors (the expected payback period went down from 15 years to 9.9 years).[90] The 'promise' of profitability was without a doubt kept.[91]

Importantly, the fact that the Czech Republic was careful to avoid retroactive regulatory changes seems important for the FET determination. While the *WA/ICW/Photovoltaik/Voltaic* Tribunal did not explicitly discuss this element, others did. For example, the *Antaris vs Spain* Tribunal considered the lack of retroactivity as decisive for its finding on the FET claims (notably in the context of legitimate expectations).[92] The *RREEF vs Spain* Tribunal commented on the importance of retroactivity in this context:

> There is, however, one aspect of the case, on which the Arbitral Tribunal has no hesitation to find that the Respondent acted in breach of its obligation to respect the principle of stability which, as recalled above, is a required obligation under the ECT, in that the challenged measures are partly retroactive.[93]

Other tribunals stressed the link between retroactivity and the FET violation as well. In *Hydro Energy vs Spain,* the Tribunal stated that while there is no general prohibition on retroactive legislation, such retroactivity 'may, depending on the context, be relevant to unreasonableness, breach of legitimate expectation [...]'.[94] In *Watkins vs Spain,* the Tribunal straightforwardly

90 *Photovoltaik Knopf Betriebs-GMBH vs The Czech Republic* at para 490.
91 *Photovoltaik Knopf Betriebs-GMBH vs The Czech Republic* at para 490.
92 *Antaris vs The Czech Republic,* para 430.
93 *RREEF Infrastructure vs Spain* (n 17) para 325.
94 *Hydro Energy 1 S.À R.L. and Hydroxana Sweden vs Spain* (Decision on jurisdiction, liability and directions on quantum, 9 March 2020) ICSID Case No. ARB/15/42, para 578.

evaluated the measure's retroactivity as a decisive factor in its FET analysis.[95] Other tribunals have taken a similar approach.[96]

A second notable factor in Tribunals' decisions was the impact that regulatory reforms had on the investment. In some cases, the lighter impact that the regulatory changes had on investments have allowed tribunals to show more deference to states' sovereignty. The *Blusun* Tribunal described the Italian measures as 'quite substantial, but was not in itself crippling or disabling'. The Tribunal further explains that 'the reduction in incentives was proportionately less than the reduction in the cost of photovoltaic technology', and that other important conditions (notably the length of the support scheme) have remained intact.[97] Also in cases against Spain, the far less drastic changes complained of in *Charanne* (in comparison with other cases against Spain) were considered by the Tribunal as more acceptable and compatible with the FET provision.

It should be noticed that at least in one case the Tribunal explicitly rejected the notion that the measures' severity is relevant. The *CEF Energia* Tribunal stated in this respect that

> [t]he Tribunal does not consider that the reasons adduced by the Respondent to justify the cut of the tariff granted for a 20-year period, [...] or the fact that the cut was not of such a magnitude as to render the investment unprofitable [...].[98]

While different tribunals have addressed different measures and circumstances,[99] the bottom line of the emerging jurisprudence seems to be almost entirely coherent. Reasonable, even substantial modification of RE support schemes will be accepted as complying with Art 10(1) ECT (i.e. the FET provision), at least as long as the investor's expectations were based on legislation, rather than specific contracts with the state. States' right to regulate therefore will not be trumped, not even in light of the future-facing, binding nature of FIT programmes.

On the other hand, more drastic measures – such as those discussed in most cases against Spain – will not be tolerated. There is a line that a state cannot cross and these cases provide a benchmark example of where this line is. Notably, states will do well to ensure that the profitability of investments is maintained to some extent, and, importantly, avoid retroactivity.

95 *Watkins Holding S.À.R.L et al. vs Spain* (Award, 21 January 2020) ICSID Case No. ARB/15/44, paras 564–570.

96 *Novenergia vs Spain* (n 39) para 697. *9REN v. Spain* (n 37) para 307.

97 *Blusun vs Italy* (Award 27 December 2016) ICSID Case No ARB/14/3, para 342.

98 *CEF Energia B.V. vs Italy* (n 82) para 241.

99 The *Charanne* claims did not cover some of the more impactful measures that were discussed in other cases, and the *Blusun* arbitration dealt with a different regulatory system altogether.

A significant change in circumstances will most likely be considered as a justified reason for making changes in existing regulations, especially where the impact of the new measures is not drastic and applied fairly and indiscriminately.

G *The nature of RE support schemes and the 'traditional' FET interpretation*

The result of these cases is not entirely surprising and, overall, it fits in well with traditional investment jurisprudence. The 'defenders' of the investment regime will be pleased with this result as it weakens the criticism according to which the FET provision could be understood as a stabilisation clause. Investment tribunals have repeatedly and explicitly rejected this interpretation, and in almost all cases the state's right to regulate was addressed, and balanced against other considerations.

The approach taken by most tribunals seems correct. The nature of schemes such as FIT is far more specific than 'normal' regulations; it is effectively a deal between two parties, where the terms and conditions are agreed upon in advance. The correct comparison here is not with a 'normal' regulation, which, of course, may change, but more with a specific contract that the parties are expected to follow, or even, as described by the UK's Court of Appeal, a government bond. The England and Wales' Court of Appeal described FITs and their nature in the following words[100]:

> The concept of a rate of payment fixed during the period of generation by reference to the date the installation became eligible for payment is fundamental to the Scheme. It provides an assurance as to the rate of return to an owner who has paid a capital sum prior to the installation coming into operation, subject to an adjustment in accordance with RPI. That the Scheme provides for a fixed rate of return during the period of generation is crucial to resolution of this appeal. Identification of the concept of a fixed rate does not depend upon any Explanatory Note, although those notes underline that concept. The fixed return to the owner assured by the Scheme was rightly described by Mr Grodzinski as analogous to the fixed rate of return on a Government bond.

Many of the arbitral tribunals discussed above, albeit not all, have made this distinction. At the same time, it is worth noting that tribunals have largely ignored the public importance of RE support schemes. Arguably, investments that are crucial for the delivery of public goals should enjoy a higher

100 The Secretary of State for Energy and Climate Change vs Friends of the Earth and Others [2012] EWCA Civ 28 <https://prospectlaw.co.uk/wp-content/uploads/2014/08/Court-of-Appeal-Judgment.pdf> para 40.

level of protection. The critics of investment law are fast to use these 'public goals' as reasons to demand higher levels of regulatory flexibility for states, and a reduced level of protection for private investors. It would make equal sense to apply a similar logic in the other direction, where investors' activities are promoting the very same public goals.

Regardless of this criticism, it seems that the emerging conclusion from these early cases is that investment protection rules can play, and are playing, an important role in the promotion of climate change programmes. They provide an additional layer of protection for RE-related investment, one that sits above and away from the host state's control. These cases demonstrate that even in 'developed' countries – EU Member States – this layer of protection is far from being redundant.

H The RE case-law and its 'positive' regulatory chill

Importantly, the cases discussed above could result with a certain 'positive' regulatory chill. According to the regulatory chill argument, 'policy-makers take into account potential disputes with foreign investors before they even begin to draft a policy and prioritize avoiding such disputes over the development of efficient regulation in the public interest'.[101] The regulatory chill argument is often used against the investment regime, stating that it inhibits states' ability to regulate fields such as human rights and environmental protection. The UN Special Rapporteur on Human Rights and the Environment – Professor David Boyd – has even warned (uncritically, in the author's view[102]) in his 2018 report that[103]:

> Corporations have used Investor-State dispute settlement mechanisms in investment treaties to file lawsuits seeking compensation for lost profits when climate policies are strengthened, creating regulatory chill. States should withdraw consent to arbitration or negotiate a carve-out for climate actions as a safeguard to protect against these types of lawsuits.

This argument is not entirely without basis. While this study is being written, several investors have submitted claims against states wishing to phase

101 Kyla Tienhaara 'Regulatory chill and the threat of arbitration: A view from political science' in Chester Brown and Kate Miles (eds.), *Evolution in Investment Treaty Law and Arbitration* (CUP 2011).

102 See more about this specific issue in the chapter on Environmental law HCMs in this study.

103 UNGA, *Report of the Special Rapporteur on the Issue of Human Rights Obligations Relating to the Enjoyment of a Safe, Clean, Healthy and Sustainable Environment* (2019) un Doc A/74/161 <https://undocs.org/en/A/74/161>.

out coal production,[104] arguably for the purpose of 'chilling' states from doing so. Whether these claims will be successful remains to be seen. Indeed the previous 'flagship' litigation that 'regulatory chill' advocates have clung to – the plain-packaging/anti-smoking cases – have ended with tribunals dismissing investors' arguments and supporting states' right to regulate.

As discussed by this author elsewhere,[105] it could be that in the case of RE investment the 'regulatory chill' will play a positive role, by forcing states to think twice before revoking support schemes and undermining investors' confidence in this area. Some indication of this positive role was noted in the *Antaris vs Czech Republic* Award, which cites the Explanatory Report published by the Czech government prior to the amendment of the regulatory regime. This Note explains the decision to avoid retroactive changes to the FIT programme in the following words[106]:

> Investors may prepare sufficiently in advance for amendment of the conditions for investment, which should entirely eliminate the risk of potential lawsuits against the Czech Republic related to protection of investments.

As the voices against the field of international investment law (and international law more generally) are heard as loudly as ever, it is important to recognise the complexity of the picture. Not all foreign investors are 'evil' and some, as in the cases discussed above, are in fact playing a crucial role in achieving worthy public objectives, such as those involved in the fight against climate change. Investment protection rules have a definite and valuable role to play in this reality.

III Investment promotion under IIAs[107]

Most IIAs are relying almost exclusively on investment protection provisions. Only a few IIAs also include investment promotion provisions; a fact that explains the little attention that this type of HCMs has received from authors and states alike. The UNCTAD's secretariat defined the difference

104 See for example a Notice submitted by a German investor to the Netherlands (Damian Charlton 'The Netherlands is put on notice of a treaty-based dispute' IAReporter (12 May 2020) https://www-iareporter-com.uea.idm.oclc.org/articles/netherlands-put-on-notice-of-a-treaty-based-dispute/>) and an arbitration launched against Canada (Westmoreland Mining Holdings vs Canada).

105 Avidan Kent 'Renewable energy disputes before international economic tribunals: A case for institutional 'greening'? (2015) 12(3) *Transnational Dispute Management*.

106 *Antaris vs Czech Republic* (n 92) para 414.

107 This part is based on the research conducted, and conclusions made, in Avidan Kent 'Policy coherence and the promotion of foreign direct investment in the renewable energy sector: Lessons from Europe' (2018) *The Yearbook on International Investment Law and Policy 2017* 22.

between investment protection provisions and investment promotion provisions in the following words[108]:

> Investment promotion provisions stand out as a special category in IIAs since – contrary to the treaty obligations concerning investment protection – they establish a commitment of the contracting parties to do something. While investment protection is geared to prevent contracting parties from taking certain measures – e.g. to discriminate against investors or to expropriate them without proper compensation – investment promotion goes further and demands from contracting parties to become active.

As explained in an UNCTAD report, investment promotion provisions vary in their nature and shape[109] and could therefore fall under different categories of HCMs. A cursory survey of examples may therefore be useful.

A Information-based investment promotion provisions

To begin with, some IIAs instruct the parties to facilitate the flow of information so as to enable investors to make better-informed decisions. Such provisions could include instructions to share information regarding investment opportunities, or, more generally, to increase transparency with respect to the host state's regulatory environment.[110] Regarding the usefulness of such measures, a study prepared by the Australian Treasury states that[111]:

> [t]ransparency is a critical input to the investment decision. This has been well documented by business surveys. Lack of transparency and predictability often tops the list of concerns expressed by foreign investors. On the flip side, access to relevant information is often cited as a powerful incentive to invest.

It is true that the obligation to provide information will traditionally fall on the host country, rather than investors' home country. However, as explained by Sauvant, this need not always be the case[112]:

108 UNCTAD, *Investment Promotion Provisions in International Investment Agreements* (UNCTAD 2008) 6.

109 Ibid., 5.

110 See surveys of investment promotion provisions in UNCTAD (n 108) 14; Anthony VanDuzer et al., *Integrating Sustainable Development into International Investment Agreements: A Guide for Developing Country Negotiators* (Commonwealth Secretariat 2013), Chapter 8. Many of the examples reviewed in this chapter are from these helpful surveys.

111 Roy Nixon 'Transparency obligations in international investment agreements' <https://treasury.gov.au/sites/default/files/2019-03/International_investment_agreements.pdf> 57.

112 Karl Sauvant 'China, the G20 and the international investment regime' (2016) 24(4) *China & the World Economy* 72, 86.

> Home countries, too, can increase transparency. From the perspective of investors, transparency is not only important as far as host countries are concerned, but also as regards support offered to outward investors by their home countries. Thus, home countries could commit to making comprehensive information available to their outward investors on the various measures they have in place, both to support and restrict OFDI.

Examples of IIAs transparency-related obligations (admittedly imposed mostly on host countries) do exist. Notably, a general obligation to provide accessible information on domestic laws, regulations and administrative proceedings is not uncommon in international investment law.[113] The Dutch Model BIT provides a general instruction to provide information on relevant 'laws, regulations, judicial decisions, procedures and administrative rulings', including a clarification that '[w]henever possible, such instruments will be made available through the internet in English'.[114] Another example can be found in the US-Rwanda BIT, which provides a detailed list of rules and decisions that local authorities will have to make public, including an instruction to establish a 'Contact Point' that will facilitate the communication of such information.[115] The Brazil-Guyana BIT as well instructs the establishment of a 'Focal Point' that will provide such information to investors from the other Party.[116]

Far less common are obligations to ensure the flow of information regarding investment opportunities. One rare example is found in the Nigeria-Morocco BIT, which instructs that '[t]he Parties shall disseminate among the relevant business sectors, general information on investment, regulatory frameworks and business opportunities in the territory of the other Party'.[117] Similarly, Schedule II of the Common Market for Eastern and Southern Africa (COMESA) investment agreement (title 'promotion and awareness programme') prescribes that the Parties shall '[o]rganise joint investment promotion activities e.g. seminars, workshops, inbound familiarization tours for investors from capital exporting countries, joint promotion of specific projects with active business sector participation'.[118] Art 24 of the Association of Southeast Asian Nations (ASEAN) Comprehensive Investment Agreement is yet another example. The Parties here intend to increase foreign investment by 'organising and supporting the organisation of various briefings and seminars on investment opportunities and on investment

113 See examples in UNCTAD, *Transparency* (UNCTAD 2012) <https://unctad.org/en/PublicationsLibrary/unctaddiaeia2011d6_en.pdf>.
114 Art 4 Dutch Model BIT (2019).
115 Art 10 and 11 of the US-Rwanda BIT (2012).
116 Art 19(4)(e) Brazil-Guyana BIT (2018).
117 Art 5(6) of the Nigeria-Morocco BIT (not yet in force).
118 Schedule II of the Investment Agreement for the COMESA Common Investment Area (signed 23 May 2007).

laws, regulations and policies; and conducting exchanges on other issues of mutual concern relating to investment promotion'.[119]

i Technical assistance and capacity building

Another type of investment promotion HCMs is a requirement to provide host states with technical assistance in order to improve their capacity to host investments. By doing so, home states effectively facilitate the investment process by creating a better investment environment in the host states. One interesting example is found in Art 25 of the Nigeria-Morocco BIT ('Assistance and Facilitation for Foreign Investment'), in which it is stated that the

> Home State should assist the Host State in the promotion and facilitation of foreign investment in particular by their own investors [...] Such assistance may include, inter alia: Capacity building with respect to Host State agencies and programs on investment promotion and facilitation.[120]

Other relevant examples appear in the ASEAN-New Zealand-Australia FTA's investment chapter, in which the parties agree to assist certain ASEAN Member States, *inter alia* by providing technical assistance to strengthen their capacity to attract investment, indirectly making it easier for their own investors to invest in these countries.[121] The Southern African Development Community (SADC) Model BIT prescribes that in the context of investment promotion, 'State Party X shall provide technical assistance to State Party B in the implementation of this Article',[122] an obligation that applies (according to the annexed Commentary) only 'in a developed/developing State context'.[123] In South/South relations, the Commentary explains that exchanges of best practices might be more appropriate.[124]

ii Financial support and investment guarantees

A rather rare investment promotion HCM involves the granting of financial incentives by home countries, for investments made in the other party's territory. Admittedly, not many examples of such provisions exist, and where

119 ASEAN Comprehensive Investment Agreement (signed 26 February 2009, entered into force 24 February 2012), Art 24.

120 Art 25, Nigeria-Morocco BIT (2016).

121 See: Agreement Establishing the ASEAN-Australia-New Zealand Free Trade Area (signed 27 February 2009, entered into force 10 January 2010), Investment Chapter Art 15. A somewhat similar provision can also be found in the ASEAN Comprehensive Investment Agreement, Art 23.

122 Art 23.4 Southern African Development Community Model BIT (SADC Model BIT).

123 SADC Model BIT (n 122) 44.

124 Ibid.

these are found they tend to be vague and non-obligatory in nature. One interesting example is Article 34 of the European Free Trade Association (EFTA)-Egypt Agreement, according to which the EFTA states 'declare their readiness' to provide financial support to Egypt, in order to facilitate investment by EFTA's investors in this country.[125]

Other somewhat similar examples are found in EU Association Agreements. These agreements, however, are very different from IIAs, and can better be seen as complementary to such agreements. Some of these agreements (e.g. the EU-Egypt Association Agreement) even explicitly call for the conclusion of IIAs as a means to create a more 'conducive' legal environment for investors.[126] The EU-Egypt Association Agreement mentions the possibility that investment in Egypt will be promoted through[127]:

- appropriate means of identifying investment opportunities and information channels on investment regulations,
- roviding information on European investment regimes (such as technical assistance, *direct financial support, fiscal incentives and investment insurance*) related to outward investments and enhancing the possibility for Egypt to benefit from them,
- legal environment conducive to investment between the two Parties, where appropriate through the conclusion by the Member States and Egypt of investment protection agreements, and agreements to prevent double taxation,
- *examining the creation of joint ventures*, especially for SMEs and, when appropriate, the conclusion of agreements between the Member States and Egypt, – *establishing mechanisms for encouraging and promoting investments.*

Admittedly, it is not clear from the language of the agreement who will be paying for such support, and whether it could be considered as a HCM.

The EU-Economic Community of West African States (ECOWAS) and the West African Economic and Monetary Union (UEMOA) Economic Partnership Agreement provide more interesting examples. While this agreement cannot be regarded as an IIA, it is possible that it will be linked to such an agreement in the future: this agreement's 'rendezvous' clause explicitly instructs the Parties to discuss the possibility of concluding an IIA in the future. Moreover, the promotion of investment is one of this agreement's objectives,[128] where specific sectors are explicitly

125 See: Free Trade Agreement between the European Free Trade Association (EFTA) and Egypt (signed 27 January 2007, entered into force 1 September 2008), Art 34.

126 See European Communities (EC)-Egypt Association Agreement (signed 25 June 2001, entered into force 1 June 2004), Art 46.

127 Ibid.

128 See: Economic Partnership Agreement between the West African States, the Economic Community of West African States (ECOWAS) and the West African Economic and Monetary Union (UEMOA) and the EU, Preamble, Art 1.2(d), and Art 56(3)(f).

targeted.[129] Importantly for this section, Article 61 of this agreement establishes a 'regional fund'. This fund 'shall be the main financing instrument of the EPA Development Programme. It shall be a preferred instrument for channelling support from the European Union and its Member States'.[130]

Another interesting example is found in the South African Development Community (SADC) Model BIT. According to Article 23.3 of this Model BIT[131]:

> The State Parties may provide Investment financing and Investment guarantee facilities for Investors from their State into the territory of the other State Party. Such facilities shall, if used, promote compliance with the obligations of Investors set forth in this Agreement.

The reader will note that the language in this agreement is extremely vague, and indeed the Commentary on this Model BIT describes this provision as 'a minimal first step in this direction'.[132]

IV International investment agreements and PCD

Much have been written about the interaction between investment protection rules and other policy-areas such as trade, human rights, environmental protection and more.[133] The majority of the literature on this topic is 'defensive' in nature, i.e. it is focussed on host states' efforts to protect their regulatory flexibility and secure other interests. As such, this literature is important, albeit outside the scope of this study. More relevant to this study are the efforts of home countries to ensure policy coherence for development (PCD) in their HCMs and in the support granted to their home investors.

The following section will therefore focus mostly on investment promotion measures, and other obligations that can be regarded as under the control of investors' home country.

A Investment promotion and PCD: not enough, too simple and too little

A 2008 UNCTAD study states with respect to investment promotion measures in IIAs[134]:

> A major deficiency in most IIAs dealing with investment promotion is their lack of specificity. They often provide no details on the

129 Ibid. Mentioned sectors are Agriculture (see Art 48(2)), Fisheries (Art 49(1)(c)).
130 Ibid., Art 61.
131 Art 23.3. SADC Model BIT.
132 *SADC Model Bilateral Investment Treaty Template: With Commentary* (SADC 2012) 44.
133 See for example Freya Baetens (ed.) *Investment Law within International Law* (CUP 2013).
134 UNCTAD (n 108) 21.

> pre-conditions and modalities of the investment promotion activities agreed upon, nor do they include any follow-up mechanism that would allow monitoring as to what extent the promotion measures have been put in place and are effective.

This description was accurate in 2008, and it is accurate also at the time of writing. This lack of detail is accompanied with (or probably the result of) a more general lack of interest in investment promotion in the context of IIAs. Most IIAs do not include such provisions to begin with, and when they do, as stated above, these provisions are not of great use. This state of affairs is probably the result of several factors. Notably, home-states may not be interested to include rules that will legally require them to invest resources, or restrict the operation of their national institutions. Another possibility is that IIAs are notoriously regarded as not more than an easy 'photo-op' for states' leaders, and indeed many of them are a result of a mere 'copy-pasting' exercise, without much thought and deliberation.[135]

Whatever the reason is, it is clear that this 'over-simplicity' is damaging in the context of PCD, especially for RE-related FDI. To begin with, most IIAs are addressing only a limited type of investors' rights, notably rights that are relevant at the post-investment stage. The *crucial* pre-investment stage is traditionally ignored, and very little is done to address the difficulties faced by investors at this stage. As discussed above, on very few occasions pre-investment HCMs such as direct financial contributions, insurance and information, were, indeed, mentioned in IIAs. These examples are too few and far between, and are usually expressed in a vague and non-binding language.

This 'over-simplicity' is damaging also when one considers PCD and links with other HCMs. Recent reforms in treaty-making practices were mostly reactive in nature; they followed the criticism expressed by NGOs and academics, which focussed almost exclusively on IIAs restrictions on states' right to regulate. The focus on the right to regulate is without a doubt important, but it led to modifications that were almost exclusively 'defensive' in nature: the addition of carve-outs, exceptions and clarifications[136] with no attention granted to investment promotion provisions. Despite their potential in promoting investment and overcoming a range of investment barriers, investment promotion provisions remained as simplistic and rare as they have ever been.

States will do well to invest more thought into the development of investment promotion provisions. IIAs currently address only a very limited range

135 This possibility could perhaps explain the lack of investment promotion provisions in older treaties, as new treaties are without a doubt scrutinised and not as easy to conclude as perhaps they once were.

136 See review in Markus Gehring and Avidan Kent 'Sustainable development and IIAs: From objective to practice' in Armand de Mestral and Celine Levesque (eds.) *Improving International Investment Agreements* (Routledge 2013).

of investment barriers, namely, political risks. If the objective of *promoting* foreign investment is to be achieved, these treaties should include, or otherwise be linked to, a variety of complementary instruments. This process is occurring where states are addressing their IIAs as part and parcel with other areas, notably by including them as chapters in wider arrangements such as trade and development agreements. While this trend is certainly on the rise, this is not yet the case for most countries.

B The way forward: IIAs as complementary tools, instead of standalone regimes

This chapter reviewed the role that both investment protection and promotion provisions can play, and already are playing, in the wider effort to promote foreign investment in the RE sector in the developing world. With respect to investment protection provisions, the conclusion is rather straightforward: while certain modifications could be made in order to make these provisions more sustainable,[137] it is clear that these provisions are useful. Notably, they provide investors with assurances that address certain investment barriers, primarily those that are related to policy risks.

The case of investment promotion provisions is somewhat more difficult to assess, notably due to the lack of examples and experience. It is argued in this chapter that states should explore the possibility of increasing the role for investment promotion provisions, according to relevant investment barriers (that is other than policy risks) and preferably on a case-by-case approach, in light of the context and circumstances of each negotiations partner.

A different possibility would be to re-think the role and shape of IIAs. In other chapters in this study, this author has called for the incorporation of investment treaties into other international law HCMs, including from the fields of trade, environment, development and insurance. These other HCMs in fact act as investment promotion instruments, albeit with far more detail (and effectiveness) than any IIA. Instead of designing and adding a wide range of complex RE-related investment promotion provisions to IIAs, states could simply add *investment protection* rules to other international law HCMs. This should not be difficult to do: investment protections provisions are relatively simple and untechnical, and their generic nature makes them easily adjustable to different contexts and subject-areas. As mentioned in other chapters in this study, some states are indeed following this concept. For example, some state-backed insurance providers are noting the existence of IIAs as an additional layer of protection. Wider free trade agreements, which include investment chapters, are effectively doing the same. In other international law HCMs – notably those that are based

137 Gehring and Kent (n 136).

on environmental and development agreements – the use of investment protection 'add-ons' is much more limited (or even non-existent).[138]

This study's conclusions will call for the creation of a wide, holistic agreement that will include all types of RE-related international law HCMs coherently, and in line with the notion of PCD. As a first step in this direction, which will be easy and uncomplicated to implement, states should consider the use of investment protection treaties as 'add-ons' for other international law HCMs.

138 This is despite calls made by academics to do so. See for example Freya Baetens 'The Kyoto Protocol in investor-state arbitration: Reconciling climate change and investment protection objectives' in Marie-Claire Cordonier Segger, Markus Gehring and Andrew Newcombe (eds.) *Sustainable Development in World Investment Law* (Kluwer 2011) 707; Anatole Boute 'Combating climate change through investment arbitration' (2012) 35(3) *The Fordham International Law Journal* 614.

4 International insurance law

De-risking sustainable investments

I Introduction

Another type of international law Home Country Measures (HCM) addressed in this study are insurance mechanisms; defined by the UNCTAD Secretariat as a distinctive group of HCMs.[1] As already discussed elsewhere in this study, the risks that are embedded in investment in the developing world are perceived as high, a fact that makes investments in these regions more expensive. One way to de-risk these investments would be an increased reliance on insurance mechanisms.[2]

The role that public insurance mechanisms can play in the efforts to de-risk sustainable investment was recognised by the international community. The Addis Ababa Action Agenda on Financing for Development is stating in this respect[3]:

> We will consider the use of insurance, investment guarantees, including through the Multilateral Investment Guarantee Agency, and new financial instruments to incentivize foreign direct investment to developing countries, particularly least developed countries, landlocked developing countries, small island developing States and countries in conflict and post conflict situations.

Of more relevance to the topic of this study, it has been argued by some that state-backed guarantees are 'the most effective leverage for at-scale policies' in the Renewable Energy sector.[4] The Intergovernmental Panel on Climate Change (IPCC) further emphasised on this topic:

1 UNCTAD, *Home Country Measures* (UNCTAD 2001) 10.

2 Schmidt (2014) 238.

3 Addis Ababa Action Agenda of the Third International Conference on Financing for Development (UN 2015) <https://sustainabledevelopment.un.org/index.php?page=view&type=400&nr=2051&menu=35>, para 45.

4 Dipak Dasgupta, Jean-Charles Hourade and Seyni Nafo, *A Climate Finance Initiative: To Achieve the Paris Agreement and Strengthen Sustainable Development* (2019) <http://www2.centre-cired.fr/IMG/pdf/rapport_gicf.pdf> 3–4.

DOI: 10.4324/9780429277207-5

> private insurers, and by means of public mechanisms, [that] can help to partially or fully reduce the exposure of investors to political risk, exchange rate fluctuations, business interruption, shortfalls in output, delays or damage during fabrication, construction, and operation of a product, project, and company.[5]

Despite the clear need to integrate insurance mechanisms into international laws and policies, international climate law has addressed these only rarely, and in passing. The UNFCCC instructs that the Parties

> shall give full consideration to what actions are necessary under the Convention, including actions related to funding, insurance and the transfer of technology, to meet the specific needs and concerns of developing country Parties arising from the adverse effects of climate change and/or the impact of the implementation of response measures [...].[6]

The Paris Agreement addresses insurance-related tools only partially, and in a very limited context (notably in the context of Loss and Damage).[7]

The following part provides a brief explanation concerning the role of insurance mechanisms in the context of this study.

II The role of insurance home country measures in the efforts to promote RE-related investment

The high risks embedded in renewable energy-related investments in the Global South (see Chapter 1 of this study) are increasing the cost of finance.[8] While the price of renewable energy (RE) technologies has been steadily reducing over the years, the cost of finance has remained significantly high due to a combination of factors. These include, *inter alia*, the high upfront investment that is required in new RE projects; the perceived risks that are associated with investment in the RE sector,[9] as well as the risks that are associated with investment in the developing world.[10] These

5 IPCC Sujata Gupta et al (eds) 'Chapter 16: Cross-cutting Investment and Finance Issues' in *Climate Change 2014: Mitigation of Climate Change* (IPCC 2014) 1226–1227.

6 See Article 4 (8) of the UNFCCC.

7 See Article 8 of the Paris Agreement.

8 Jan Christoph Steckel and Michael Jacob 'The role of financing cost and de-risking strategies for clean energy investment' (2018) 155 *International Economics* 19; UNEP (n 12) 36.

9 See for example in Gianleo Frisari, Morgan Hervé-Mignucci, Valerio Micale and Federico Mazza 'Risk gaps: A map of risk mitigation instruments for clean investments' (2013) *Climate Policy Initiative* <http://climatepolicyinitiative.org/wp-content/uploads/2013/01/Risk-Gaps-A-Map-of-Risk-Mitigation-Instruments-for-Clean-Investments.pdf>; Nadine Gatzert and Thomas Kosub 'Risks and risk management of renewable energy projects: The case of onshore and offshore wind parks' (2016) 60 *Renewable and Sustainable Energy Review* 982; Dasgupta, Hourade and Nafo (n 4) 5.

10 See a review of these risks in Chapter 1.

additional costs of finance can be dramatic: according to a 2013 United Nations Development Programme (UNDP) estimate, the differences in the cost of finance may amount to an additional 40% in comparison with equivalent projects in the developed world, or, alternatively, in comparison with equivalent gas-based projects in the developing world.[11]

The high cost of finance does not only make finance more difficult to come by; it also increases investors' expected revenues from projects,[12] implying higher costs for consumers and states.[13] This increase in costs for consumers/host countries stands in contrast to a number of Sustainable Development Goals (SDGs), notably SDG 7 that instructs states to provide *affordable*, sustainable and reliable energy for their citizens. Other SDGs that are adversely affected by the high cost of finance in this context are SDG 1 ('no poverty'), SDG 8 ('economic growth'), SDG 9 ('industry, innovation and infrastructure'), SDG 13 ('climate action') and more. The importance of de-risking projects is therefore clear.[14]

Where insurance policies are providing sufficient and affordable cover, their impact has a positive 'domino' effect: de-risking investment via insurance mechanisms will reduce the cost of finance, and consequently the projected revenues that investors will expect.[15]

Different types of policies[16] and insurance products are available in the market. While this study will address primarily political risk insurance (PRI), it is important to stress that other instruments exist, and that their role should not be neglected. One important example is the World Bank's Guarantees Program. The Guarantees Program offers both project-based and policy-based guarantees, some of which are of use to RE investors (notably payment guarantees,[17] regulatory risks and political risks). The World Bank describes this Program as a 'powerful catalyst to attract private-sector

11 Oliver Waissbein et al., *De-risking Renewable Energy Investment. A Framework to Support Policymakers in Selecting Public Instruments to Promote Renewable Energy Investment in Developing Countries* (2013 UNDP) 34.

12 UNEP, *De-risking Renewable Energy Investment: A Framework to Support Policy-Makers in Selective Public Instruments to Promote Renewable Energy Investment in Developing Countries* (UNEP 2015) 35.

13 Where states are deciding to subsidise these additional costs, these higher costs will be imposed on them, rather than directly on consumers.

14 IPCC, *Climate Change 2014: Mitigation of Climate Change. Contribution of Working Group III to the Fifth Assessment Report of the Intergovernmental Panel on Climate Change* (CUP 2014) 1211.

15 World Bank, *A Sure Path to Sustainable Solar: Solar Deployment Guidelines* (World Bank 2019) <http://pubdocs.worldbank.org/en/155991570472678574/A-Sure-Path-to-Sustainable-Solar-Guidelines.pdf> 39.

16 Steckel and Jacob (n 8); UNEP (n 12); Rogerio Studart and Kevin Gallagher 'Guaranteeing sustainable infrastructure' (2018) 255 *International Economics* 84.

17 See for example an RE project in which these guarantees have supported solar investment <http://pubdocs.worldbank.org/en/208821518200595153/Briefs-Guarantees-Zambia ScalingSolar.pdf>.

investments',[18] and claims that over the past 20 years it has mobilised $42 billion in commercial capital and private investment.[19]

More recently (2019), the World Bank launched the Solar Risk Mitigation Initiative (SRMI). The SRMI's declared purpose is to support developing countries' efforts to leverage private investment in RE. This programme supports only solar projects for the time being, but is expected to also cover wind projects in the future. The SRMI is still modest in scope[20] and as such, it is far too early to assess its effectiveness.[21] However, at the same time, this Program represents an ambitious model, with special relevance for this study. A World Bank report framed the SRMI's raison d'être in the following words[22]:

> Yet, why are there so few privately owned solar and wind projects in developing countries? What do countries need to do to see a significant deployment of private investments in VRE, deployments that would be aligned with national needs while also being affordable?

While the programmes discussed in this study are certainly useful, academics have argued in the past that more targeted programmes are necessary for ensuring and increasing sustainable and climate-friendly investment. These academics have made their own proposals in order to bridge this gap. For example, Studart and Gallagher[23] have proposed the establishment of a global green guarantee fund, which will be funded through international pledges and will provide guarantees 'to promote a project pipeline, to monitor the development and impact of projects, and to mitigate the risks involved in sustainable infrastructure investments'.[24] While this proposal is very brief and short on details, these authors do stress that funds will be conditioned by recipient states' commitment to stability (e.g. unreasonable changes to support schemes) and the mitigation of political risks.[25] These conditions imply links to investment treaties (which these authors do not mention), and a greater need for Policy Coherence for Development (PCD), as discussed below.

18 The World Bank 'Guarantees program' <https://www.worldbank.org/en/programs/guarantees-program>.

19 Ibid.

20 Only USD 500 million are allocated to this Program at the time of writing.

21 One early indicator (October 2019) was France and the Netherlands' commitment to invest USD 100 million towards RE projects in Africa, including USD 55 million as guarantees for private investors in sub-Saharan Africa. IISD 'Mitigating finance update: Every project counts to help achieve SDGs 7, 13' *IISD News* <https://sdg.iisd.org/news/mitigation-finance-update-every-project-counts-to-help-achieve-sdgs-7-13/>.

22 World Bank, *A Sure Path to Sustainable Solar: Solar Deployment Guidelines* (n 15) 7.

23 Studart and Gallagher (n 16).

24 Studart and Gallagher (n 16) 90.

25 Studart and Gallagher (n 16) 90.

Another proposal was made by Dasgupta et al. They envision the establishment of a new mechanism, 'backed by sovereign guarantees [that] should initially be restricted to projects contributing to host countries' NDCs, thus linking them closely to those countries' enabling policies'.[26] The purpose of this envisioned mechanism is to reduce the risks that are typical to the RE sector, *inter alia* the high upfront capital cost; new technologies-related risks; the long duration of the investment, and uncertain carbon prices.[27] This mechanism will be based on the principle of common but differentiated responsibilities, and will be initially funded through the 'USD 100 billion per year' Copenhagen financial pledges.[28]

A different model was proposed by Steckel and Jakob,[29] who likewise recommend a list of measures for the purpose of de-risking investment in the RE sector. Many of these measures are focussed on domestic frameworks (e.g. increased transparency and inclusivity), and are beyond the scope of this study. Some measures, however, could be implemented on the international sphere, notably their call for improved policy coherence,[30] and the use of different forms of subsidies for addressing investors' risk/returns ratio.[31] Importantly for this chapter, these authors are stressing the significance of Political Risk Insurances, which will be discussed in more detail below. These authors emphasise the current lack of utilisation of insurance for the purpose of leveraging of international investment in RE:

> For climate-related investments, however, to date policy risk insurance constitutes only a relatively small fraction of lending by multi-lateral development banks. Whereas in 2015 direct loans for climate-related investments accounted for about US$ 40 bn, all risk management instruments combined (which include policy risk insurance as a subcategory) amounted to only roughly US$ 1.5 bn (CPI, 2015). By creating the enabling environment to leverage private finance, providing more support for risk management tools could turn out to be a more effective approach than direct project-based support.[32]

The proposals made above demonstrate the need to develop public mechanisms that will de-risk investment in the RE sector. This chapter will address the role of insurance mechanisms and their potential in increasing investment in the RE sector. For reasons related to space and focus, this chapter will concentrate mostly on Political Risk Insurance: an instrument

26 Dasgupta, Hourade and Nafo (n 4) 5.
27 Dasgupta, Hourade and Nafo (n 4) 5.
28 Dasgupta, Hourade and Nafo (n 4) 24.
29 Steckel and Jacob (n 8) 22–26.
30 Steckel and Jacob (n 8) 22–26.
31 Steckel and Jacob (n 8) 25; IPCC (n 14), 1211.
32 Steckel and Jacob (n 8) 26.

that has already found its way into the world of international law HCMs, and one that fits well with other instruments reviewed in this study.

III PRI-related instruments

Political Risk Insurances (PRIs) provide investors with protection from political events that adversely affect their investment. The types of events protected by PRI vary according to specific policies, but most will protect against events such as expropriation, wars and different forms of political violence.[33] Certain authors have rightfully pointed at the overlap between PRIs and the protection provided by International Investment Agreements (IIAs).[34] There are however notable differences, including the cost of protection (IIA's protection is free while PRIs are expensive), length and cost of recovery (PRIs are much faster, and do not require litigation-related expenses) and with respect to substantive and procedural protection (wider protection is granted via IIAs).[35] More differences include the ease of recovery (usually far easier via PRIs, especially where states are refusing to pay arbitration awards), and the fact that – as discussed below – strong institutional actors are assisting in the early resolution of disputes and providing their backing to investors insured via public PRIs.

Both private and public entities are offering insurance policies for potential investors. Private insurers such as Lloyds, Marsh or Chubb are offering a variety of PRIs, covering risks ranging from consequential financial loss due to political actions, political violence and politically motivated confiscation of property. Private insurance policies, however, are often limited in their scope and coverage, and in most cases are very expensive.[36]

33 MIGA defines 'political risks' as risks that

> are associated with government actions which deny or restrict the right of an investor/owner (i) to use or benefit from his/her assets; or (ii) which reduce the value of the firm. Political risks include war, revolutions, government seizure of property and actions to restrict the movement of profits or other revenues from within a country.

34 Pieter Bekker and Akiko Ogawa 'The impact of Bilateral Investment Treaty (BIT) proliferation on demand for investment insurance: Reassessing political risks after the 'BIT Bang'' (2013)28(2) *ICSID Review* 314.

35 Bekker and Ogawa (n 34). These authors have commented on the additional substantive protection provided by BITs:

> In addition to providing protection from expropriation, fair and equitable treatment and currency convertibility, BITs typically also offer substantive and procedural guarantees such as protection against arbitrary or discriminatory measures, national treatment, most-favoured-nation treatment, full protection and security, the minimum standard of protection under international law, and access to arbitration.

36 Nadine Gatzert and Thomas Kosub 'Risks and risk management of renewable energy projects: The case of onshore and offshore wind parks' (2016) 60 *Renewable and Sustainable Energy Reviews* 982, 993; Hannah Mayer, *Political Risk Insurance and Its Effectiveness in Supporting Private Sector Investment in Fragile States* (The LSE-Oxford Commission

Public insurance instruments[37] provide a useful alternative in this respect. As explained by Mayer, public insurance providers are able to assume more risks as they are less motivated by profits; they offer longer coverage and are far more affordable.[38] Frisari et al further add that public insurers are offering 'a more stable presence in the industry'[39] and, as will be discussed below, can rely on their political power (notably the Multilateral Investment Guarantee Agency (MIGA)) to effectively resolve disputes between insured investors and host states.[40] Peinhardt and Allee commented on the importance of this last advantage in the context of the Overseas Private Investment Corporation (OPIC, now DFC), the US public insurer[41]:

> Where a potential claim arises, a public insurer like OPIC can rely on connections with US Departments of Treasury and State (and others) to help resolve the problem before a claim occurs, or to pressure host states to provide compensation afterwards. Private insurers, even those with strong lobbying histories, will never be able to marshal the power of multiple government agencies to resolve a potential claim. And this advantage extends past the claim stage to the recovery of assets paid. Historically, OPIC has maintained a recovery rate of over 90%, meaning that it eventually recovers almost all of the money it pays out in claims.

While there are many types of public insurance providers, in line with this book's theme, this study will address mostly international law-based public insurance mechanisms and the role that these can play in the de-risking of investments and the reduction of investment costs.

---

on State Fragility, Growth and Development 2018) 9. The reader should note that the claim according to which private PRI is more expensive is disputed by private insurers, see for example Clint Peinhardt and Todd Allee 'Political risk insurance as dispute resolution' (2016) 7 *Journal of International Dispute Settlement* 206, 208–209.

37 It should be noted that there are different models of 'public' insurance providers, where some agencies are operating under close governmental supervision and others enjoy greater autonomy. See more details in Kathryn Gordon 'Investment guarantees and political risk insurance: Institutions, incentives and development' (2008) OECD Investment Policy Perspectives 2008 <http://www.oecd.org/daf/fin/insurance/44230805.pdf> 97.

38 Mayer (n 36).

39 Peinhardt and Allee (n 36) 209.

40 Gianleo Frisari et al. argue that

> Political, policy and social risks are exogenous and therefore difficult for private parties – who have limited ability to enforce compliance on public authorities – to manage. Thus, following the OECD classification (OECD, 2008), we note that the public sector (at national or international levels) would be better positioned to enforce compliance and lower the probability of their occurrence.

Frisari, Hervé-Mignucci, Micale and Mazza (n 9) 13.

41 Peinhardt and Allee (n 36) 209.

IV Insurance mechanisms in bilateral agreements

The first type of international law-based insurance mechanisms is grounded in (or at least related to) bilateral treaties. Insurance mechanisms are mentioned in different types of bilateral agreements, whether as part of trade/investment agreements or within specialised designated insurance agreements.

A References in 'non-insurance' agreements

A review of environmental, development, trade and investment treaties did not reveal many references to insurance mechanisms, although, notably, a few references are made in investment treaties. For example, Art 25 of the Nigeria-Morocco Bilateral Investment Treaty (BIT) instructs that '[t]he Home State should assist the Host State in the promotion and facilitation of foreign investment in particular by their own investors. [...] Such assistance may include, inter alia [...] Insurance programs based on commercial principles'.[42] It should be stated that, at the time of writing, this treaty has not yet come into force, and it is difficult to know whether this vague instruction will be executed and how.

Another example is Art 23.3 of the Southern African Development Community (SADC) Model BIT, according to which

> [t]he State Parties may provide Investment financing and Investment guarantee facilities for Investors from their State into the territory of the other State Party. Such facilities shall, if used, promote compliance with the obligations of Investors set forth in this Agreement.[43]

Interestingly, the Commentary to the SADC Model BIT states that the use of such HCMs allows Home Countries further control over the process of investment, and may therefore also result in increased accountability and responsibility[44]:

> The article allows Home States to require that its investors who seek to make an investment under the treaty comply with the obligations contained herein as a condition of State financing or insuring of the investment. This gives some specific responsibility to the Home State for the conduct of its investors where governmental facilities are being used to support the investor.

42 Art 25 Nigeria-Morocco BIT (2016).

43 Art 23 SADC Model BIT (2012).

44 *SADC Model Bilateral Investment Treaty Template: With Commentary* (SADC 2012) <https://www.iisd.org/sites/default/files/meterial/6th_annual_forum_sadc_model_template.pdf> 44.

It is evident from the examples mentioned above (or more accurately, from the lack of examples) that, despite clear relevance to investment promotion and protection, the few references made to insurance mechanisms in investment and trade agreements are mostly vague and are made in a non-committed manner. These references could be seen more as future aspirations than actual substantive legal provisions: in their current form, they simply offer nothing to investors. Such references fall under the first category of UNCTAD's classification – policy positions – defined as 'positive in tone but vague in specific commitments'.[45] The limited value that these provisions offer to investors, accompanied with their scarcity (especially in the case of North-South agreements), is enough to conclude that these examples cannot be considered as meaningful.

At the same time, these provisions are not entirely without utility. They place the foundations for more serious attempts to incorporate insurance mechanisms into investment/trade agreements. The fact that treaty negotiators are linking investment protection with insurance is encouraging on its own. One may hope that a certain continuation will follow, and more sophisticated mechanisms will, indeed, be added to these agreements.

B Insurance agreements

Bilateral insurance treaties are concluded in order to facilitate and regulate the operation of national agencies (e.g. export credit agencies) that aim to support local investors' activity in foreign markets. These agencies provide investment guarantees, including insurances against investment risks (*inter alia* PRIs). Sauvant et al define the operation of these national agencies as HCMs for obvious reasons[46]: their main task is to support the operation of home states' investors in foreign countries through the reduction of risks that are associated with their investment.

Many of these agencies are operating without any grounding in international agreements. The operation of such agencies will be addressed only briefly in this study, as they do not qualify as *international law* HCMs. As discussed elsewhere in this book, the fact that these agencies are not operating as international law-based HCMs is problematic from the perspective of development: at least on the face of it, they are designed to represent only the commercial interests of investment exporting states, while host states have very little say in the design of their rules of operation.[47]

45 UNCTAD (n 1) 8.

46 Karl Sauvant et al. 'Trends in FDI, home country measures and competitive neutrality' in Andrea Bjorklund (ed.) *Yearbook on International Investment Law & Policy 2012–2013* (OUP 2014) 16–18. <http://ccsi.columbia.edu/files/2014/03/Yb-12-13-ch.-1-8-Nov-13-stand-alone-final-for-website.pdf.>

47 Mayer indeed straightforwardly commented that these agencies are 'commercially-focused as their primary aim is to promote their countries' exports rather than conduct development work'. Mayer (n 36) 12.

It should be noted that, at least on the face of it, international law could mitigate the one-sidedness that is implied from the operation of these agencies. As briefly discussed below, some of these agencies are explicitly bound by international rules that dictate certain levels of responsibility towards other states' needs and interests. For the purpose of this part, however, the focus will be on investment promotion agencies whose operation is based on specific international insurance agreements.

One notable example of an agency that is working based on international agreements is the Overseas Private Investment Corporation (OPIC), now merged into the US International Development Finance Cooperation (DFC). The OPIC/DFC is a US government agency, whose main role is to support outward FDI through loans and guarantees. Importantly for this part, the OPIC/DFC is also providing insurances, including against political risks. This agency offers these insurances due to the lack of a viable, affordable alternative on the private markets. Indeed, according to OPIC's eligibility criteria, this agency will insure projects 'where the investor is unable to find private political risk insurance at acceptable rates or terms'.[48]

Importantly, the DFC now offers insurance against a variety of political risks, including expropriation and other forms of 'unlawful' government interference, *inter alia* the impairment of contracts and the forced negotiation of contract terms and conditions. These types of cover are useful for investors in the RE sector, where the risk of expropriation is evermore present. The OPIC/DFC's insurances are also useful in cases in which support schemes' benefits are guaranteed via contracts. It should be noted, however, that, as recent examples demonstrate (e.g. Spain, the Czech Republic), support schemes could be guaranteed also via regulation. Invoking DFC's insurances in such cases could be more complicated.

As stated above, the operation of this agency is supported through a line of international agreements with states in which supported investors operate. These agreements could be seen as international law HCMs: their aim is *inter alia*

> to encourage economic activity [in the territory of the US treaty partner in any given agreement, AK] which promote the development of the economic resources and productive capacities of the [US' treaty partner in any given agreement, AK] and to provide for investment insurance (including reinsurance) and guarantees which are backed in whole or in part by the credit or public monies of the United States of America.[49]

48 OPIC 'Political risk insurance: Eligibility' <https://www.opic.gov/content/eligibility>.

49 See for example preamble to *Investment Incentive Agreement between the Government of the United States of America and the Government of the Democratic and Popular Republic of Algeria*.

On the face of it, as is the case in other international law HCMs, the interests of both investment exporting and importing states are recognised in these treaty objectives.[50] In practice, however, these agreements are fairly light in terms of commitments and despite their grand, development-related objectives, they are addressing almost exclusively the operation of OPIC. In essence, they include instructions such as the exclusion of OPIC from local regulation, its exemption from host country's taxation, or dispute resolution. These agreements' efficiency (as HCMs) is therefore very limited.

C References in states' PRI providers to international HCMs

As stated above, not many non-insurance treaties refer to insurance mechanisms. The picture, however, is more complex when one evaluates references made by national insurance agencies to other international law instruments, notably international investment agreements (IIAs). An Organisation for Economic Co-operation and Development (OECD) study found that[51]:

> A non-representative review of policies of some State-backed PRI providers suggests that some institutions take IIAs into account for insurability and pricing, especially State-backed insurers in countries that have large treaty networks: The French and German investment insurance providers explicitly mention the presence of an IIA among the aspects considered for insurance; an IIA is a sufficient – but not a necessary – condition to establish 'adequate legal protection'. MIGA follows a similar approach. Political risk insurers from other countries, including some countries with large treaty networks – Finland or Italy, for instance – do not explicitly mention the presence of an IIA as a factor for insurance against non-commercial risk.

IIAs are not the only international law mechanisms that are addressed by state-backed insurance agencies. As will be discussed in greater detail

50 Another example of text from such treaties that also emphasise the interests of the host state includes the treaty concluded between the US and Jordan, in which it is stated that the parties are 'affirming their common desire to encourage economic activity in Hashemite Kingdom of Jordan that promote the development of the economic resources and productive capacities of the Hashemite Kingdom Jordan' and 'Acknowledging that diverse forms of assistance and support are important to the economic revitalization of the Hashemite Kingdom of Jordan'. See in the preamble to the *Investment Incentive Agreement between the Government of the United States of America and the Government of the Hashemite Kingdom of Jordan*.

51 Joachim Pohl 'Societal benefits and costs of international investment agreements: A critical review of aspects and available empirical evidence' (2018) OECD Working Papers on International Investment 2018/01 <https://www.oecd-ilibrary.org/docserver/e5f85c3d-en.pdf?expires=1595612668&id=id&accname=guest&checksum=42E5FEE471A0CD3DA7213416B106E95D> 35.

below, many state-backed PRI agencies also refer to CSR-related instruments, notably: the OECD's Guidelines for Multinational Enterprises; Recommendation on Common Approaches on Environment and Officially Supported Export Credits; the OECD Anti-Bribery Convention, and the Action Statement on Bribery and Officially Supported Export Credits.[52]

D Case-study: OPIC (DFC) and PCD

As stated above, there are many examples of national agencies offering PRI to investors in order to promote foreign investment. The operation of many of these agencies is somewhat in line with the notion of PCD. For example, most OECD states' agencies are imposing environmental and societal standards on their investors.[53] One interesting example that is worth recalling concerns the operation of the US' OPIC/DFC which, unlike other national agencies, is supported by international agreements and thus could be linked to the topic of this study.[54]

The operation of OPIC/DFC has supported many RE-related investments. According to a study prepared by Reuters, OPIC's support for investments in solar projects in the year 2017 exceeded $250 million – a rise of more than 100% since 2016, but still significantly lower than its support in previous 'Obama' years (2013–2015).[55] A cursory glance at OPIC's list of supported projects reveals that OPIC supported dozens of RE-related projects, a significant number of which are located in Africa.[56]

The DFC's mission is certainly in line with the notion of PCD. This agency has adopted an Environmental and Social Policy (January 2020), aiming, *inter alia*,

> to provide countries a robust alternative to state directed investments by authoritarian governments and United States strategic competitors using best practices with respect to transparency and environmental

52 Gordon (n 37) 100.

53 Gordon (n 37) 100.

54 Unfortunately, at the time of writing, the DFC's new website does not include detailed important information on policies and standards. The author therefore assumes that many of OPIC's policies were transplanted 'as is' into the DFC's operation, and are therefore still valid. Even if this is no longer the case (again, the author was not able to locate alternative policies), the OPIC presents a useful case study, and as such deserves special discussion. Where available, the DFC's more recent policies will be discussed as well.

55 See online <http://fingfx.thomsonreuters.com/gfx/editorcharts/USA-SOLAR-OPIC/0H0010KS08B/index.html>.

56 See a list of OPIC-supported projects <https://www.opic.gov/opic-action/all-project-descriptions>.

and social safeguards, which take into account the debt sustainability of partner countries.[57]

As stated above, the OPIC/DFC is operating based on a variety of bilateral treaties, concluded between the US and the countries in which OPIC/DFC-supported projects are operating.[58] Many of the 150 agreements currently concluded between the US and other states concerning the operation of OPIC (none mentions the DFC) are fairly old, often pre-dating the conclusion of the UNFCCC and the Rio Declaration, making it unlikely to find references to these agreements or, more broadly, to issues such as climate change and renewable energy. A cursory review of more recent agreements (those concluded after the year 2000) reveals that these agreements are often following the same, very basic (and brief) model, briefly mentioning the host state's development in the preamble but with nothing on the promotion of public goals such as climate change abatement or the promotion of renewable energies.

With respect to investment in renewable energies, the DFC's Environmental and Social Policy states that the DFC will ensure that its supported projects 'are environmentally and socially sustainable' and '[a]re compatible with low and no-carbon economic development'.[59] Furthermore, the DFC's Environmental and Social Policy has a section on 'Climate Change and Renewable Energy', which is discussed below.[60]

One relevant element in the OPIC's operation are their targeted guidelines for investments in wind, geothermal, biofuels, solar and hydro energy.[61] These guidelines are almost entirely 'defensive' in nature, i.e. they are aiming to ensure that RE projects in foreign countries will not harm these countries' environment.[62] They essentially require assessment of the projects' impacts on issues such as habitats, socio-cultural issues, disposal of equipment at the end of its lifetime and more. These guidelines are certainly important; ensuring investors' positive environmental behaviour is one of the features that any policy aimed at the promotion of RE-related

57 DFC 'Environmental and social policy and procedures' (January 2020) <https://www.dfc.gov/sites/default/files/media/documents/DFC_ESPP_012020.pdf> para 1.2.

58 See the list of bilateral treaties signed by the US concerning the operation of OPIC online <https://www.opic.gov/content/bilateral-agreements>.

59 DFC 'Environmental and social policy and procedures' (n 57) para 1.3.

60 DFC 'Environmental and social policy and procedures' (n 57) Part 8.3–8.11.

61 See online <https://www.opic.gov/doing-business-us/OPIC-policies/renewable-guidelines>.

62 For example, OPIC's Environmental Guidance Renewable Energy – Solar Projects states that

> [t]hese guidelines present potential environmental and social issues associated with solar PV energy projects, how OPIC may consider each of these issues when screening projects, applicable guidelines and standards, recommended measures to mitigate impacts, information needed to review a project and monitoring recommendations.
> <https://www.opic.gov/sites/default/files/files/Solar-Jan31-to-OPIC.pdf>

investment should include. Such measures, however, are not HCMs, but more of complementary policies that are aiming to ensure that outward investments in RE are sustainable.

Beyond their specific guidelines on RE-related investment, OPIC also has an 'Environmental and Social Policy Statement' (OPIC's Statement), that was issued in 2017.[63] Unlike equivalent statements made by other national agencies (e.g. UKEF, see below), OPIC's Statement almost entirely ignores international laws or guidelines.[64] While statements are not usually regarded as binding legal instruments, the purpose of this Statement is to provide 'Applicants notice of the general environmental and social requirements that OPIC applies in evaluating prospective projects and monitoring on-going OPIC-supported projects'.[65] The Statement further 'outlines how OPIC will put into practice its commitment to the development goals' and 'will ensure through its processes that projects receiving OPIC support [...] [a]re environmentally and socially sustainable'.[66] Moreover, with respect to climate change and renewables (Part 8 of the Statement), it is stated that '[t]he requirements contained within this section apply to all projects supported through OPIC insurance, reinsurance, direct loans, or investment guaranties, including support through Financial Intermediaries'.[67] In other words, the Statement is addressed as a binding legal mechanism.

Elsewhere in this Statement it is noted that OPIC has committed to cut emissions from its supported projects by 50% by the year 2023 (a 'cap' that was inherited by the DFC[68]). As part of its efforts to reach this target, OPIC has adopted a cap for its entire covered operation; each new project is allocated a portion of this emission cap for its own operation. Projects that rely on RE, or that achieve a significant reduction in their emissions, are allocated an emission cap of zero (for the purpose of OPIC's overall cap); an incentive that allows OPIC to exclude them from its overall cap. While this incentive is granted for energy users, when it comes to energy producers, it is likely that RE-generation investments will qualify as 'zero emissions', especially where these are replacing existing power plants and thus lead to significant reductions.

In summary, the example of OPIC is interesting for several reasons. Most importantly, OPIC's commitments to the promotion of RE investment in the developing world are noticeable in this agency's practice. These commitments, however, are domestic in essence and are not reflected in the

63 OPIC 'Environmental and social statement' (2017) <https://www.opic.gov/sites/default/files/files/final%20revised%20ESPS%2001132017(1).pdf>.

64 A few references do exist in OPIC's Statement, for example to the International Labour Organization's Declaration on Fundamental Principles and Rights at Work.

65 OPIC Statement (n 63) 1.1.

66 OPIC Statement (n 63) 1.3.

67 OPIC Statement (n 63) 8.1.

68 DFC 'Environmental and social policy and procedures' (n 57) para 1.3.

international treaties based on which OPIC is operating across the globe. It could be that the US' longstanding refusal to internationally bind itself on matters such as climate change has something to do with the lack of reference to international law. However, the fact that climate change is not the only public objective to be ignored by US' insurance treaties (in fact public goals, except for a vague reference to economic development in the preamble, are categorically ignored in these treaties) suggests an alternative explanation. It could be that current US insurance treaties are simply at the same 'evolutionary' stage as the old-generation (pre-2000) international investment agreements, i.e. copy-pasted, extremely short agreements that are replicated without much debate and thought. It is hoped that just as in the case of investment law, increased public and academic scrutiny will shed more light on this area in international law and will lead to its evolution, including in terms of connectivity with other areas of international law.

V Multilateral insurance agreements

A second category of international law-based insurance providers are multilateral, public insurers. These insurers qualify as international law-based HCMs; they are formed by states via international law, and aim to increase foreign investment through mitigation of risks. The main actor in this group of insurers is without doubt the World Bank's investment insurance branch – the Multilateral Investment Guarantee Agency (MIGA). MIGA's operation is based on international law, notably the Convention establishing the Multilateral Investment Guarantee Agency (MIGA Convention), which was signed by 181 states.

The MIGA Convention's objective is[69]:

> [T]o encourage the flow of investments for productive purposes among member countries, and in particular to developing member countries, thus supplementing the activities of the International Bank for Reconstruction and Development (hereinafter referred to as the Bank), the International Finance Corporation and other international development finance institutions.

This objective is certainly in line with the concept of PCD: MIGA is a supplementing tool, one that is aimed at enhancing the efforts of other international laws and institutions and to maximise their operation. MIGA's mandate in this respect is further enforced by Art 35, which instructs MIGA to cooperate with the UN and other relevant IGOs.

69 Art 2 of the MIGA CONVENTION <https://www.miga.org/sites/default/files/archive/Documents/MIGA%20Convention%20February%202016.pdf>.

MIGA provides guarantees to private sector investors operating in the developing world, notably in the form of PRIs. The risks covered by the MIGA Convention are non-commercial in nature,[70] including restrictions on currency transfer, direct/indirect expropriation, the breach of contract by host states (where recourse to the courts is unavailable), and wars and civil disturbances.[71] This list is somewhat open, as Art 11(b) states that

> the Board, by special majority, may approve the extension of coverage under this Article to specific non-commercial risks other than those referred to in Section (a) above, but in no case to the risk of devaluation or depreciation of currency.

Although MIGA's coverage is currently limited to USD 250 million, it can cover higher sums through reinsurance.[72]

While the focus of MIGAs operation is usually on PRI, it should be noted that this agency's efforts to promote investment are not limited to this tool only. For example, Art 23 of the MIGA Convention ('Investment Promotion') also mentions activities such as research, dissemination of information on investment opportunities ('with a view to improving the environment for foreign investment flows to such [developing] countries') and technical assistance. Importantly, MIGA also facilitates the amicable settlement of disputes between insured investors and their host states. MIGA's affiliation with the World Bank is especially useful in this context as, according to its website, it provides an 'umbrella of deterrence against government actions that could disrupt insured investments'.[73] While it is difficult to evaluate the credibility of MIGA's claims in this respect (MIGA does not publish data on such disputes), some authors are claiming that MIGA is indeed playing an important role in this respect.[74] Another successful feature of MIGA's dispute settlement services is its mediation facility. Again, for understandable reasons, MIGA does not make information on such disputes available to the public.

In light of the surge in disputes between investors and host states concerning the operation of RE support schemes (notably the retrospective cancellation of allegedly promised benefits), it is hard to ignore MIGA's dispute settlement capabilities, or its deterrence 'powers' in this context. The

70 MIGA Convention (n 69), Art 2(a).

71 Ibid., Art 11.

72 MIGA 'Investment guarantee guide' (MIGA 2015) <https://www.miga.org/Documents/IGGenglish.pdf>.

73 MIGA 'Dispute resolution' <https://www.miga.org/investment-guarantees/dispute-resolution>.

74 Axel Drehel et al. 'Membership has its privileges – The effect of membership in international organizations on FDI' (2015) 66 *World Development* 346; Paul Collier 'Attracting international private finance for African infrastructure' (2014) 1 *Journal of African Trade* 37, 42.

backing of entities such as the World Bank could be useful for recovering payments debts,[75] alleviating at least some of the risks involved in investment in the Global South.[76]

Authors are envisioning further roles for MIGA in the developing world, possibly in collaboration with other institutions. Collier, for example, recommends that World Bank funds should be directed towards covering the cost of insurance premiums for infrastructure projects in Africa.[77] Such projects could include investment in the RE sector. While MIGA insures mostly political risks, Collier also mentions the need to insure other risky elements in investment in the developing world, including currency exchange risks.[78] This idea could also be developed in the more specific context of the RE sector, which requires the alleviation of its own specific risks.[79]

Lastly, while this chapter will not elaborate on other multilateral insurance providers, it is important to stress that MIGA is not the only player in this field. Notably, there are also other, regional, multilateral public insurance providers. One example is the African Trade Insurance Agency (ATIA); an international agency that was established through a treaty[80] between primarily Eastern African states.[81] Interestingly, with respect to the promotion of investment in renewables, ATIA is collaborating with other multilateral public insurers. For example, in collaboration with the European Investment Bank (the EU's landing arm – and another multilateral public insurer), ATIA has significantly increased its insurance capacity and expertise.[82] This collaboration, as stated above, is linked to the attainment of SDG 7 and will support only investment projects that are, *inter alia*, 'aligned with at least one of the following objectives: energy access, energy efficiency, as well as renewable energy'.[83]

A MIGA and the promotion of RE-related projects in the developing world

The MIGA Convention does not include references to environmental protection, renewable energies or climate change. At least with respect to climate change and renewable energies, this omission is understandable – the

75 Paul Collier (n 74) 42.
76 Gianleo Frisari et al. (n 40) 12.
77 Paul Collier (n 74) 41.
78 Paul Collier (n 74) 41.
79 See sources cited in FN 9 in this chapter.
80 Agreement Establishing the African Trade Insurance Agency (2000) <http://www.ati-aca.org/wp-content/uploads/2019/07/ATI-Treaty-2017-Edition.pdf>.
81 For a full list of members, see online <http://www.ati-aca.org/investor-information/current-members/>.
82 See online: ATIA.
83 See in Implementation Guidelines for EU SE4All funding, Art 1.2(1) (Eligibility Criteria) <http://www.ati-aca.org/wp-content/uploads/2017/09/SE4All-EU-Guidelines.pdf>.

Convention was concluded in the early 1980s, before these issues came to prominence. This state of affairs was somewhat corrected with the conclusion of MIGA's Policy on Environmental and Social Sustainability (2007, and a new edition adopted in 2013).[84] MIGA's Environmental and Social Policy opens with the following statement[85]:

> The Multilateral Investment Guarantee Agency (MIGA) strives for positive development outcomes in the private sector projects for which it provides guarantee support. MIGA believes that an important component of achieving positive development outcomes is the environmental and social sustainability of these projects

MIGA was criticised in the past for not doing enough for promoting RE-related investment. Dasgupta *et al* mention in this respect: 'between 2004–2015, all major MDBs, including MIGA, approved a total of only USD 40 billion in project guarantees, representing only 4.4% of their total financing'.[86] In recent years, however, this situation is changing. MIGA has itself pledged to address climate change, making it one of its 'strategic priorities',[87] and stating that 'it will support innovative investments services to climate-friendly solutions'.[88] More specifically concerning the RE sector, MIGA states that its 'support for low-carbon economic development is one dimension of a balanced approach to development, including supporting access to modern, clean, and reliable energy services'.[89]

MIGA's commitment to renewables is indeed materialising into actions. In 2017, MIGA's mother institution – the World Bank – announced that it will stop financing upstream oil and gas projects[90] (albeit this is an obligation that the Bank was criticised for not respecting[91]). In 2018, MIGA mobilised 'a record-breaking $3 billion in private sector co-financing for climate-related activities'.[92] Its 2018–2020 Strategy includes '[a]n expanded commitment to climate change mitigation and adaptation,

84 MIGA 'Policy on environmental and social sustainability' (MIGA 2013) <https://www.miga.org/documents/Policy_Environmental_Social_Sustainability.pdf>.

85 MIGA 'Policy on environmental and social sustainability' (n 84).

86 Dasgupta, Hourade and Nafo (n 4) 21.

87 MIGA 'Policy on environmental and social sustainability' (n 84) para 10.

88 MIGA 'Policy on environmental and social sustainability' (n 84) para 10.

89 MIGA 'Policy on environmental and social sustainability' (n 84) para 11.

90 See report <https://www.theguardian.com/business/2017/dec/12/uk-banks-join-multinationals-pledge-come-clean-climate-change-risks-mark-carney>.

91 See for example here <https://www.desmog.co.uk/2019/04/15/world-bank-gave-21-billion-support-fossil-fuels-over-four-years-report>.

92 MIGA '2018 annual report' <https://www.miga.org/sites/default/files/2018-11/MIGA%20 2018%20Annual%20Report_0.pdf> 19.

targeting 28% of new issuance related to climate change mitigation or adaptation in 2020'.[93]

The implementation of MIGA's Environmental and Social Policy is delivered via MIGA's Environmental and Social Performance Standards (2013),[94] and supported by this Institution's environmental and social due diligence process.[95] Compliance is further supplemented through the work of MIGA's Office of Compliance Advisor/Ombudsman (CAO), which allows affected communities to raise concerns regarding MIGA's projects.[96]

Surprisingly, none of MIGA's eight Performance Standards directly addresses climate change. This is not to say, however, that this topic is entirely ignored. References to climate changes are made, *inter alia*, in Standard 1 (Assessment and Management of Environmental and Social Risks and Impacts), Standard 4 (Community Health, Safety, and Security) and, importantly, Standard 3 (resource efficiency and pollution prevention). Standard 3 explicitly instructs MIGA's clients to consider the reduction of GHG emissions (including the 'adoption of renewable or low carbon energy source'),[97] while large emitters are asked to quantify their emissions.[98]

B MIGA and PCD

On the face of it, it seems evident that MIGA is increasingly aligning its work with the notion of PCD. The adoption of this institution's Environmental and Social Policy and its Performance Standards both support this conclusion. Also, in practice, MIGA has expressed the political will to promote investment in RE and, as stated above, has even made a few commitments in this respect. At the same time, the lack of institutional coherence is noticeable: while MIGA is working very closely with other World Bank units, there is very little mention of any attempt to coordinate efforts with international trade instruments, and almost no mention of environmental law. While MIGA's Environmental and Social Policy and Performance Standards explicitly mention several treaties and IGOs (notably the International Labour Organization's (ILO) and several human rights treaties), none of these are related to other international law HCMs, or directly to investment in renewable energies.

93 MIGA 'Strategy and business outlook FY18-20' <https://www.miga.org/sites/default/files/2018-08/MIGA%20Strategy%20-%20Report%202018-2020.pdf> 2.

94 MIGA 'Performance standards on environmental and social sustainability' (2013) <https://www.miga.org/sites/default/files/archive/Documents/MIGA_Performance_Standards_October_2013.pdf>.

95 These include *inter alia* site visits by MIGA staff and a reporting mechanism, see: MIGA 'Understanding MIGA's environmental and social due diligence process' <https://www.miga.org/Documents/Understanding_MIGA_ES_Due_Diligence_Process.pdf>.

96 <http://www.cao-ombudsman.org/>.

97 MIGA (n 94), Standard 3 [7].

98 Ibid., Standard 3 [8].

Moreover, it seems that despite its strongly worded declarations on the topic, MIGA's role as an international law HCM could have been more meaningful. For example, a closer look at MIGA's Standards reveals that they are lacking in both vigour and specifics. Notably, there is no explicit instruction to prevent polluting projects from being insured. The only restriction is that investors will 'implement technically and *financially feasible* and *cost-effective* options to reduce project-related GHG emissions during the design and operation of the project'.[99] The terms 'financially feasible' and 'cost-effective' are opening the door for the insurance of polluting projects. Moreover, the requirement to 'reduce' GHG emissions was not quantified; a fact that takes away much of its edge.

MIGA's Standards (or other regulations) could, for example, have adopted an overall emission cap for its projects, similar to that adopted by OPIC. It could also have been decided that insured projects will have to rely on certain technological (green) standards. Another possibility would have been that insured projects will be asked to commit to purchase a certain percentage of their energy use from renewable sources, or, where RE is not available, to offset their emissions via emission trading schemes. Perhaps more simply, MIGA could refuse to insure projects that pollute above a certain level of GHG,[100] which, in fact, will be in line with the World Bank's recent commitment to stop financing upstream oil and gas projects.[101]

As an international law HCM, MIGA should also more actively incentivise RE-related investments. MIGA could do this by offering more attractive prices to RE investors, or by significantly increasing coverage, well above the standard USD 250 million. Again, MIGA could look into the experience of national investment promotion agencies (notably OPIC), in which far more targeted RE-related products indeed exist, and specific RE-guidelines are available.

VI Soft law mechanisms and indirect regulation of investment promotion/insurance agencies

As stated above, most public insurance agencies are not operating based on international law agreements and are therefore falling outside the scope of this study. It should be noted nevertheless, that international law also indirectly affects national insurance agencies. The role of soft law guidelines and principles is important in this respect.

99 Emphasis added. Ibid., Standard 3 [7].

100 Such a condition could include exceptions, such as the performance of emission 'offsetting' action (e.g. purchasing emission trading units under a certain recognisable emission trading schemes).

101 See report <https://www.theguardian.com/business/2017/dec/12/uk-banks-join-multinationals-pledge-come-clean-climate-change-risks-mark-carney>.

A The OECD common approaches

Notably, certain national insurance agencies have subjected their own operation to international soft law. The most widely adopted guidelines in this context are the OECD's Common Approaches for Officially Supported Export Credits and Environmental and Social Due Diligence ('Common Approaches').[102] In essence, the Common Approaches dictate the application of certain principles, *inter alia*, to obligations to prevent and mitigate environmental impact, promote sufficient assessment procedures, foster transparency, and more.[103] They provide instructions on the screening of projects, their assessment and review.

On the face of it, the Common Approaches are useful for the purpose of this study: they address policy coherence vis-à-vis certain areas, notably with climate change. The preamble of the Common Approaches confirms that OECD states are recognising 'the responsibility of Adherents to implement the commitments undertaken by the Parties to the United Nations Framework Convention on Climate Change'.[104] The Common Approaches 'objectives' are *inter alia*[105]:

> [to] [p]romote coherence between Adherents' policies regarding officially supported export credits, their international environmental, climate change, social and human rights policies, and their commitments under relevant international agreements and conventions, thereby contributing towards sustainable development

Also interestingly, the Common Approaches instruct that[106]:

> '[states] shall continue to build a body of experience on the application of this Recommendation through regular reporting and exchanges of information [...]', and that '[t]o facilitate the building of the body of experience and to give further consideration to climate change issues' states shall 'report the estimated annual greenhouse gas emissions from all fossil-fuel power plant projects'.

With respect to clean energies, the Common Approaches provide specific instructions on thermal and nuclear energy,[107] but, surprisingly, not for other types of RE. States here are instructed to 'tak[e] into account, where

102 OECD Common Approaches for Officially Supported Export Credits and Environmental and Social Due Diligence TAD/ECG(2016)3 <http://www.oecd.org/officialdocuments/publicdisplaydocumentpdf/?doclanguage=en&cote=tad/ecg(2016)3>.
103 OECD Common Approaches (n 102).
104 OECD Common Approaches (n 102), Preamble.
105 OECD Common Approaches (n 102) 3(i).
106 OECD Common Approaches (n 102), Para 46.
107 OECD Common Approaches (n 102), Para 47.

appropriate, the context of the low carbon growth framework of the country where the project is located, the use of best appropriate technology to reduce carbon emissions, and other recommended actions'. This instruction is naturally very important, but, as stated, it is restricted only for nuclear and thermal power plants.

The Common Approaches further instruct that: 'In undertaking this assessment, Adherents should take into account the industry sector, location and other information available relating to potential environmental and/or social impacts, before making any final commitment to provide official support'.[108] This instruction could imply that due to its positive impact on climate change, investment in renewables should be positively regarded by national agencies and impact their decision to support RE investors.

A closer look at the operation of investment promotion/export credit agencies reveals more on the application of these guidelines. A notable example in this respect is the operation of UK Export Finance (UKEF) – a UK governmental branch, providing political risk insurance for British investors abroad. The application form for acquiring a political risk insurance from UKEF specifically requires applicants to specify 'environmental and social impacts relating to the Investment/Project'.[109] The form enquires more specifically on a list of environmental issues, including climate change, water, air, ozone layer and more.[110] Based on the information provided, investors may be asked to submit more information or even a full impact assessment report.[111]

While many states' agencies are incorporating guidelines such as the Common Approaches into their operation, it is doubtful whether these steps could be regarded as HCMs. The UK's practice (as demonstrated from the review of the application form) suggests a 'defensive' interpretation of the Common Approaches; i.e. the objective is to protect host countries from adverse impacts of supported investment on their environment, and not the active promotion of home countries RE investors. A more active approach would be to incorporate the instruction to 'promote coherence' by providing RE investors with preferable terms and encourage their operation.

B Other international law instruments integrated in the operation of states' investment promotion agencies

While the Common Approaches are possibly the main soft-law guidance document in terms of impact on national agencies, it is certainly not the only one. For example, both the Equator Principles and the UN's Guiding

108 Common Approaches (n 102), Part III para 8.

109 See on UKEF's website <https://assets.publishing.service.gov.uk/government/uploads/system/uploads/attachment_data/file/648998/OII-application-form-sept-17.pdf>.

110 See Part B of the form (n 109).

111 See paragraph 5 of the form (n 109).

Principles on Business and Human Rights are mentioned in UKEF's Note on Social and Human Rights Risks and Impacts,[112] as well as in the preamble to the Common Approaches itself.

Other 'hard laws' are also mentioned in agencies' operative documents, including a variety of international human rights laws. UKEF, for example, defined the term 'human rights' based on conventions such as the Universal Declaration of Human Rights, the International Covenant on Civil and Political Rights, the International Covenant on Economic, Social and Cultural Rights, and the ILO Declaration on Fundamental Principles and Rights at Work. Similarly, the US OPIC's guidelines on investment in wind energy rely on the International Union for the Conservation of Nature's (IUCN) classifications in its own categorisation of prohibited projects, that OPIC will not fund.[113]

For the time being, it seems farfetched to claim that any of this normative background is affecting outward investment in the same manner as international HCMs do. In this sense, this discussion is beyond the scope of this study. It should nevertheless be noted that this situation may change in the future with the development of additional soft law mechanisms, and the increasingly central place that climate change is taking in policy-making. For example, it is not beyond imagination to conceive of similar guidelines that will encourage the implementation of clean sources of energy. The potential role of the SDGs as a future normative source is especially interesting in this respect. As stated above, existing soft laws are already referring to the UNFCCC and addressing clean energy, even if only in a limited manner. This discussion, therefore, could be relevant in the future.

VII HCM insurance mechanisms and PCD: conclusion

The role of international insurance law HCMs is significant. These tools can, and already are, supporting investment in RE projects around the world. Despite notable efforts to increase PCD and link these tools with other international law HCMs, very little has been achieved so far and implicit references to other relevant international law HCMs in insurance treaties are scarce.

It is especially surprising to find this lack of PCD in the (minimal) interaction between insurance mechanisms and investment treaties. As commented by Robert Ginsburg, this disconnect is startling as the MIGA and ICSID are both operating under the same umbrella of the World Bank, addressing

112 UKEF's Note on social and human rights risks and impacts <https://assets.publishing.service.gov.uk/government/uploads/system/uploads/attachment_data/file/604438/UKEF-statement-on-how-it-addresses-human-rights-march-2017.pdf>.

113 Overseas Private Investment Corporation Environmental Guidance Renewable Energy – wind projects <https://www.opic.gov/sites/default/files/files/Wind-Jan31-to-OPIC.pdf>.

risks which are fairly similar in nature[114] (IIAs were described by some as 'effectively act as free political risk insurance'[115]).

This result is somewhat unexpected, given the complementary role of insurance mechanisms, which are (arguably[116]) supposed to complete and support other efforts, and not necessarily to be relied on as leading drivers for investment. The lack of explicit PCD is surprising also given that MIGA's mandate is overtly allowing this organisation to search for compatibilities and linkages with other institutions. As stated above, the OECD Common Approaches are instructing the same. It is expected, for example, to see MIGA playing an increasing role in the UNFCCC process, and that states would connect their public insurance agencies' work more directly to their international development and climate change obligations.

A review of existing insurance agreements reveals that most of these are 'stuck' in the thinking of the 1980s, hardly surprising as most were indeed concluded during (or prior to) this period. A comparison to the old generation IIAs, where no references to public goals can be found, seems apt.

On the other hand, it seems unfair to describe an entirely bleak picture. The adoption of the OECD Common Approaches is a good step in the right direction, as well as the fact that this document, together with other OECD documents (notably the OECD Guidelines for Multinational Enterprises), was widely adopted by national agencies. It seems therefore, that even where the majority of state-backed PRI providers are operating with no grounding in international agreements, international guidelines (mostly soft law) still manage to find their way into these frameworks and add some level of policy coherence to their operation. References made by these agencies to IIAs further demonstrate that states' agencies are showing an appetite for PCD and are attempting to incorporate relevant international law HCMs into their operation. The problem, as stated, is with the concrete and meaningful application of these ideas. It would be useful in this respect to see insurance mechanisms integrated into wider processes, which are led by climate, trade and development institutions, rather than acting in isolation from these.

114 Robert Ginsburg 'Political risk insurance and bilateral investment treaties: Making the connection' (2013) 14 *The Journal of World Investment & Trade* 943.

115 Lise Johnson, Lisa Sachs, Brooke Güven and Jesse Coleman 'Costs and benefits of investment treaties: Practical considerations for states' (2018) Policy Paper Columbia Center on Sustainable Investment <https://papers.ssrn.com/sol3/papers.cfm?abstract_id=3277965>.

116 Dasgupta, Hourade and Nafo (n 4).

5 International trade law

A Swiss Army knife with not enough (political) edge

I Introduction

The following chapter addresses the role that international trade law can play in the efforts to promote investment in the renewable energy (RE) sector in the Global South. The importance of international trade rules as tools for the attainment of the Sustainable Development Goals (SDGs) is widely acknowledged. The Addis Ababa Action Agenda on Financing for Development states in this respect[1]:

> We recognize the significant potential of regional economic integration and interconnectivity to promote inclusive growth and sustainable development, and commit to strengthening regional cooperation and regional trade agreements. We will strengthen coherence and consistency among bilateral and regional trade and investment agreements, and to ensure that they are compatible with WTO rules. Regional integration can also be an important catalyst to reduce trade barriers, implement policy reforms and enable companies, including micro, small and medium-sized enterprises, to integrate into regional and global value chains. We underline the contribution trade facilitation measures can make to this end.

A statement made by 50 Member States of the World Trade Organization (WTO) on 'trade and environmental sustainability' stresses in this respect 'the important role of trade, trade policy and the multilateral trading system, including the WTO, in supporting the international community's efforts towards achieving the SDGs and international environmental commitments through a forward-looking, focused, systematic and action-oriented dialogue'.[2]

1 Addis Ababa Action Agenda of the Third International Conference on Financing for Development (UN 2015), para 87.

2 *Communication on Trade and Environmental Sustainability* (17 November 2020) WT/CTE/W/249, para 1(1).

DOI: 10.4324/9780429277207-6

In the more specific context of this study, it is clear that international trade laws are relevant for investment in the RE sector in a great number of different ways.[3] An OECD report mentions in this context the impact of trade remedies (notably anti-dumping and countervailing duties), technical barriers to trade (TBT), public procurement, subsidies, trade-related investment measures (TRIMS) and more.[4] A United Nations Conference on Trade and Development (UNCTAD) report on investment promotion discusses the importance and relevance of international trade for Foreign Direct Investment (FDI), stating that '[a]long with market size, openness to trade has been identified as one of the most reliable indicators of the attractiveness of a location for foreign investment'.[5]

The manner in which the reduction of trade barriers (notably non-tariff trade barriers[6]) on Environmental Goods and Services (EGS) supports RE-related investment is fairly straightforward.[7] While the production of green technologies is taking place mostly in the developed world (and increasingly also in China[8]), the demand is global in nature. In order to be effective, these green goods and services will have to be globally disseminated on a very large scale. This in turn implies an increased role for international trade, as well as a necessity to facilitate international trade in EGS through the reduction of trade barriers.[9]

Past experience demonstrates that international trade can play a positive role in the global efforts to disseminate green technologies. Notably, the entry of Chinese RE equipment producers into the RE global market (which

3 See for example Gary Hufbauer et al. (eds.) *The Law and Economics of a Sustainable Energy Trade Agreement* (CUP 2013).

4 OECD, *Overcoming Barriers to International Investment in Clean Energy, Green Finance and Investment* (OECD Publishing 2015) <http://dx.doi.org/10.1787/9789264227064-en> Chapter 4.

5 UNCTAD (n 108) 19.

6 According to some, the reduction of non-trade barriers is expected to be much more impactful than the mere reduction of tariffs, see: EU Commission, *Trade Sustainability Impact Assessment on the Environmental Goods Agreement: Inception Report* (European Commission, May 2015) <http://trade.ec.europa.eu/doclib/docs/2015/may/tradoc_153485.pdf> 15.

7 For a review of tariffs/non-tariffs trade barriers, see: Heiner Bucher et al. 'Trade in environmental goods and services: Opportunities and challenges' (2014) International Trade Centre Technical Paper <http://www.intracen.org/uploadedFiles/intracenorg/Content/Publications/AssetPDF/EGS%20Ecosystems%20Brief%20040914%20-%20low%20res.pdf>.

8 Maria Carvalho et al. 'Understanding the dynamics of global value chains for solar photovoltaic technologies' (2017) WIPO Economic Research Working Paper No. 40 <https://www.wipo.int/edocs/pubdocs/en/wipo_pub_econstat_wp_40.pdf>.

9 OECD, The World Bank, UN Environment, *Financing Climate Futures: Rethinking Infrastructure* (OECD Publishing 2018) <https://www.oecd-ilibrary.org/environment/financing-climate-futures_9789264308114-en> 59.

led to increased competition and a reduction in prices) was enabled through international trade in RE production equipment.[10]

The following chapter will review some of the international trade law Home Country Measures (HCMs), which could potentially support RE-related FDI in the developing world.

II Reduction of tariffs on EGS as an international law HCM

The first part of this chapter will evaluate the efforts made to reduce trade barriers on EGS. Most of these efforts are focused on the reduction/elimination of tariffs; a strategy criticised by many as limited at best. This is notably due to the fact that tariffs on most environmental goods are already low, but also because the more significant, non-tariff trade barriers are still mostly ignored.

In 2011, the members of the Asia-Pacific Economic Cooperation (APEC) organisation agreed to reduce tariffs on a list of 54 EGS to 5% or less.[11] This early commitment was voluntary in nature, and questionable in terms of ambition as the average tariffs on these 54 EGS was already low (2.6%), well under the APEC's 5% objective.[12] Nevertheless, this effort was considered by many as a relative success. To begin with, despite its voluntary nature, most APEC states have indeed implemented these commitments and tariffs on a variety of goods were reduced.[13] Second, the APEC experience also led to wider and more ambitious attempts to reduce tariffs on EGS, notably within the WTO where a group of 46 Member States are currently negotiating a far more comprehensive Plurilateral WTO Environmental Goods Agreement (EGA).[14]

The choice of Plurilateral negotiations (i.e. not all WTO members are included, only those who choose to ratify the negotiated future agreement) is interesting. On the one hand, only 46 out of the 164 WTO Member States are currently involved in the negotiations. While it is true that key states (e.g. China, the US, EU, Japan and Korea) are included in this group of negotiating states, the vast majority of the developing world is not. If this state of affairs does not change, the impact of the EGA on RE investment in the Global South will be very limited. This is important, especially in light of the

10 Carvalho et al. (n 8) 16–17.

11 See analysis and background in Rene Vossenaar, *The APEC List of Environmental Goods: An Analysis of the Outcome & Expected Impact* (2013 ICTSD) <http://www.ictsd.org/sites/default/files/downloads/2013/06/the-apec-list-of-environmental-goods.pdf>.

12 Vossenaar (n 11).

13 Rene Vossenaar, *Reducing Import Tariffs for Environmental Goods: The APEC Experience* (ICTSD 2013) <https://www.ictsd.org/sites/default/files/research/reducing_import_tariffs_for_environmental_goods_the_apec_experience.pdf> xii.

14 <https://www.wto.org/english/tratop_e/envir_e/ega_e.htm.>

fact that tariffs are problematically high mostly in the developing world,[15] and not necessarily in the West. On the other hand, there is no doubt that even in its current form, the green industry will benefit from a successful conclusion of the EGA; a result that will enforce the RE sector as a whole and also indirectly improve its ability to enter into less lucrative markets.

At the time of writing, the WTO's EGA negotiations are at a halt as the negotiating states seem to be unable to agree on a list of goods that will be covered by the EGA.[16] The commercially based divisions are striking, and even non-controversial, obvious environmental goods such as bicycles have been opposed to[17] (former European Trade Commissioner Cecilia Malmstrom bizarrely explained that the EU had 'quite cheap bicycles already'[18]).

A EGA: an insufficient start

While it is important to recreate a momentum and bring the EGA-negotiating parties back to the table, it is equally crucial to send a message that in its current form the EGA is expected to be insufficient. Notably, the lists of EGS that are currently negotiated have been criticised as inadequate. Jacob and Møller have stated that tariffs on complementary goods (i.e. goods that are not environmental per se, but are nevertheless necessary for the completion of RE projects) are hardly included despite being significantly high – as high as 10%–20% in some cases.[19] de Melo and Solleder further criticise the EGA's concentration on tariffs. These authors mention in this respect the importance of non-tariff barriers, including local-content requirements, anti-dumping measures, intellectual property (IP) protection, labelling and standards.[20] Notably, the omission of trade in services is especially damaging, given that a wide variety of services are complementary to the trade in goods in this sector.[21]

It is important to note the role and position of developing countries with respect to the EGA negotiations. In short, developing countries are not taking part in this effort (Costa Rica being the one exception). de Melo and

15 Jaime de Melo and Mariana Vijil 'Barriers to trade in environmental goods and environmental services: How important are they? How much progress at reducing them?' (2014) FEEM Working Paper No. 36.2014, <http://papers.ssrn.com/sol3/papers.cfm?abstract_id=2428192> 23.

16 Jaime de Melo and Jean-Marc Solleder 'The EGA negotiations: Why they are important, why they are stalled, and challenges ahead' (2018) FERDI Working Paper 236 <http://www.ferdi.fr/sites/www.ferdi.fr/files/publication/fichiers/wp236_melo-solleder_0.pdf> 3.

17 de Melo and Solleder (n 16) 4.

18 Tom Miles 'EU blames China for WTO environmental talks collapse' (4 December 2016) *Reuters* <https://www.reuters.com/article/us-trade-environment/eu-blames-china-for-wto-environmental-trade-talks-collapse-idUSKBN13T0MX>.

19 Jacob and Møller (n 25) 20.

20 de Melo and Solleder (n 16) 9–11.

21 Jacob and Møller (n 25) 32–37.

Solleder explain that this state of affairs is as a result of wider geo-political divisions and tensions, notably developing countries disappointment with the United Nations Framework Convention on Climate Change (UNFCCC) process and the limited contribution that the Global North has made so far in this context. Vossenaar proposes that developing countries should receive special and differential treatment in order to convince and incentivise them to join the EGA negotiations.[22] Such a proposal, however, could be counter-productive and damaging to the efforts to promote foreign investment in the RE sector, as it will prioritise the establishment of inefficient/non-existent domestic industry over proven knowledge/capital rich foreign firms.

B EGS: future pathways

While the WTO example seems to have reached a dead end, the idea behind the EGA negotiations is not without merit. To begin with, even if limited in scope and coverage, the EGA could serve as a starting point, creating an institutional framework that will facilitate more significant reductions in the future. Indeed, the EU foresees the EGA as a 'living agreement', which will allow states to add more goods in the future, as well as, importantly, trade in services.[23]

An interesting, albeit much more complicated layer that could be added to the EGA involves trade in RE, i.e. the electricity itself, rather than the goods and services that are used to produce it. Facilitating international trade in RE will, without a doubt, increase the potential market as well as the incentive to invest. Brandi recommends in this respect that the WTO Member States should consider an arrangement in the spirit of the Energy Charter Treaty, in order to liberalise this aspect as well.[24]

Furthermore, while, as mentioned above, tariffs on EGS are mostly low, this is not always the case. Even among APEC states where EGS' tariff reduction commitments were made, and by and large implemented, tariffs on RE equipment are still above the average for industrial goods.[25] In APEC's least developed countries, the average tariff on such goods is 4.73% (in comparison to a 2.1% average for other industrial goods), which is still significant.[26] Others also add that tariff negotiations under the WTO, which could

22 Rene Vossenaar 'Identifying products with climate and development benefits for an environmental goods agreement' (2014) ICTSD Issue Paper No. 19, 29.

23 See for example this press release <http://trade.ec.europa.eu/doclib/press/index.cfm?id=1116>.

24 Clara Brandi, Trade 'elements in countries' climate contributions under the Paris Agreement' (2017) *ICTSD Climate and Energy* <https://euagenda.eu/upload/publications/untitled-81229-ea.pdf> 26.

25 Arun Jacob and Anders Møller 'Policy landscape of trade in environmental goods and services' (2017) ARTNeT Working Paper Series No. 166 <https://www.unescap.org/sites/default/files/AWP%20No.%20166.pdf> 20.

26 Jacob and Møller (n 25) 20.

result in reduced *bound* tariffs (rather than applied tariffs) will help reduce financial uncertainties and increase investor confidence.[27]

Lastly, the idea behind the EGA could be improved, and translated into bilateral and regional frameworks. Development agreements, for example, could also include (alongside other guarantees and commitments) more targeted and suitable lists of liberalised EGS (see for example New Zealand-Taiwan Free Trade Agreement or FTA). These lists could be based on specific, predictable projects envisioned by the negotiating states at the time of negotiations. The more limited bilateral/regional frameworks could also allow for the establishment of joint committees to approve new RE projects' lists for future projects. These envisioned committees would be able to approve more than just EGS, but also necessary complementary equipment and relevant services, either per project or on a permanent basis.[28]

III Trade-related intellectual property rules as international law HCMs

Another topic that deserves attention in the examination of trade rules and their impact on RE-related investment is trade-related IP rules. The protection of IP rights under trade and investment treaties is, and always has always been, a contentious topic. This is especially true in the context of development. For example, many developing countries view the WTO's Agreement on Trade-Related Aspects of Intellectual Property Rights (TRIPS) as a one-sided agreement, which aims to promote the economic interests of the West, while adversely affecting developing countries' interests.[29] These concerns are not entirely baseless as indeed the TRIPS negotiations were initiated and advocated by developed nations, in order to ensure that their own industries benefitted from access to wider markets. In this sense, TRIPS is a classic international law-based HCM – an assurance (or even an incentive) that was negotiated by investors' home states in order to facilitate and promote outward investment and trade. Indeed, most will agree today that these rules were not accepted as part of developing countries' efforts to attract

27 de Melo and Vijil (n 15) 11.

28 A fairly limited version of these committees could be found in agreements such as the Trans-Pacific Partnership (TPP) and the USMCA. For example, the parties to the TPP established an Environment Committee, as well as Contact Points (Article 20.19) that are acting as a forum for discussion and are entrusted with the overseeing of the implementation of the TPP's environment chapter. The Committee is also (vaguely) entrusted with continuing discussions on the liberalisation of EGS (see Article 20.19 TPP, and Article 24.24(3) of the USMCA).

29 Notably public health, culture and the promotion of infant industries. See review in Poppy Winanti and Alasdair Young 'Complying with unwelcome rules? Developing countries and the TRIPs agreement' (2009) 2 *The Indian Journal of International Economic Law* 52.

investment (i.e. host country measures), but rather insisted upon by investment exporting countries as a tool for supporting exporting industries.

The relevance of IP protection for RE-related technology development and dissemination is clear,[30] even if its impact is still debated. While a significant part of the transformation into a greener economy is expected to take place in the Global South, RE technologies are produced mostly in the North. According to Dussaux et al., 75% of the patented low-carbon inventions are registered in either the US, the EU, Japan or South Korea.[31] It is not unreasonable to assume that patent owners will be reluctant to operate in jurisdictions in which their IP rights are not sufficiently protected. Indeed, research on this topic has been mostly supportive of this claim, although, importantly, as discussed below, a consensus does not seem to exist.[32]

As stated by Ockwell et al.,[33] the impact that IP protection rules may have on RE projects could be understood by two, rather conflicting narratives. The first narrative is that strong IP protection is necessary for the promotion of RE-related FDI in the developing world. As stated above, research on this topic has been somewhat supportive of this claim. According to the second narrative, however, the great public importance of low-carbon-related IP (equated to pharmaceuticals) should lead to the reduction (or even the expropriation) of IP protection. The objective, according to the supporters of the second narrative, is to make RE technologies as freely available as possible, for the benefit of society as a whole. This second narrative has gained some traction in UNFCCC negotiations,[34] as well as within the WTO.

It is important to note that the true impact of IP protection in this context is debated not only by states, but also by researchers. While some studies have shown that IP protection does not significantly inhibit RE-related technology transfer,[35] others claim that IP rights in fact inhibit the innovation

30 IRENA, *Intellectual Property Rights: The Role of Patents in Renewable Energy Technology Innovation* (IRENA 2013) <https://www.irena.org/-/media/Files/IRENA/Agency/Publication/2013/Intellectual_Property_Rights.pdf>.

31 Damien Dussaux et al. 'Intellectual property rights protection and the international transfer of low-carbon technologies' (2018) Centre for Climate Change Economics and Policy Working Paper No. 323 <http://www.lse.ac.uk/GranthamInstitute/wp-content/uploads/2018/01/Working-Paper-288-Dussaux-et-al.pdf> 2.

32 Eduardo Porter 'Lifting patent barrier to new drugs and energy resources' (12 April 2016) *New York Times* <https://www.nytimes.com/2016/04/13/business/economy/lifting-thepatent-barrier-tonew-drugs-and-energy-sources.html>.

33 David Ockwell et al. 'Intellectual property rights and low carbon technology transfer: Conflicting discourses of diffusion and development' (2010) 20 *Global Environmental Change* 729.

34 Ockwell et al. (n 33) 730.

35 Ockwell et al. (n 33) 733–736.

and dissemination of green technologies.[36] More nuanced conclusions were perhaps presented by Dussaux et al.[37] These authors measured the impact of IP protection on international trade and investment in RE technologies. They discovered a significant positive impact, especially in FDI related to solar, heating, lighting and clean vehicles.[38] In the more specific case of non-OECD countries, the impact of IP protection is significantly noticeable in FDI related to solar and wind technologies, as well as clean vehicles.[39] At the same time, a certain negative impact was found with respect to trade in goods related to solar technologies.[40] These authors admit that their results are far from homogeneous and adjustments to any IP rules should be made on a case-by-case basis.

A TRIPS/IP chapters, renewables and PCD

At least on the face of it, the TRIPS agreement was supposed to deliver when it comes to the transfer of green technologies to the Global South. Technology transfer into the developing world is indeed one of the objectives of TRIPS.[41] Developing nations, however, are not content with the implementation of this obligation,[42] and indeed, as discussed earlier in this study, most RE-related investment is taking place mostly within the developed world.

Certain states are attempting to address this problem within the WTO. Notably, Ecuador floated the link between RE and TRIPS in a communique from 2013.[43] Unsurprisingly, Ecuador chose to emphasise the narrative according to which IP rights inhibit the dissemination of climate-related technologies. Ecuador reminded the WTO Member States that the UNFCCC enshrines the principle of common but differentiated responsibilities, and that the application of this principle requires the transfer of technologies to the developing world. This country, therefore, called for the reduction of IP protection on RE technologies and for an increased reliance on

36 Kilian Raiser et al. 'Corporatization of the climate? Innovation, intellectual property rights and patents for climate change mitigation' (2017) 27 *Energy Research & Social Science* 1; Bronwyn H. Hall and Christian Helmers 'The role of patent protection in (clean/green) technology transfer' (2010) NBER Working Paper No. 16323 <https://www.nber.org/papers/w16323.pdf>.

37 Dussaux et al. (n 31).

38 Dussaux et al. (n 31) 34.

39 Dussaux et al. (n 31) 34.

40 Dussaux et al. (n 31) 34.

41 See Art 7, 8(2) and 66 of the TRIPS.

42 Jayashree Watal and Leticia Caminero 'Least-developed countries, transfer of technology and the TRIPS agreement' (2017) WTO Staff Working Paper ERSD-2018-01 <https://www.wto.org/english/res_e/reser_e/wpaps_e.htm>.

43 Communication from Ecuador 'Contribution of intellectual property to facilitating the transfer of environmentally rational technology' WTO Doc IP/C/W/585 <https://www.wto.org/english/tratop_e/trips_e/cchange_e.htm>.

flexibility mechanisms, to allow developing countries an easier access to RE technologies.

Ecuador's communication started a heated debate. Unsurprisingly, developing countries mostly supported Ecuador's position (Art 7 TRIPS being repeatedly referred to[44]), while developed countries objected,[45] for all the reasons stated above. Like many other contentious topics, this political impasse has not yet been breached.

Attempts to address this matter are also taking place within other forums, notably the UNFCCC. The Member States of the UNFCCC (which are vastly the same as the Member States of the WTO) have agreed on a Technology Mechanism that is supposed to achieve a similar result – the promotion of international trade and investment in RE-related goods and services. Under the UNFCCC framework, however, there are hardly any references to IP protection or to TRIPS. A cursory review of documents related to the Poznan Strategic Programme on Technology Transfer (and the programme's implementation[46]) is supportive of this conclusion. As explained by Dussaux et al.,[47] the reason for this omission is, once again, the political divisions between the Member States on the role of IP protection in this context.

Beyond TRIPS and the UNFCCC, IP protection is addressed in numerous preferential trade arrangements (PTAs).[48] The manner in which PTAs approach IP protection varies to a great extent.[49] Seuba reviewed the universe of PTAs and confirmed that indeed the more 'onerous' IP-related instructions are found in PTAs in which at least one of the parties is a developed country, i.e. with significant economic interests in safeguarding IP

44 Art 7 TRIPS states that 'The protection and enforcement of intellectual property rights should contribute to the promotion of technological innovation and to the transfer and dissemination of technology…'.

45 See extract from minutes of meeting of the council for trade-related aspects of intellectual property rights held on 25–26 February 2014, WTO Doc IP/C/M/75/Add.1 <https://www.wto.org/english/tratop_e/trips_e/cchange_e.htm>, as well as other minutes referred to here <https://www.wto.org/english/tratop_e/trips_e/cchange_e.htm>.

46 See for example a report issued by the GEF, *Implementing the Strategic and Long-term Programs on Technology Transfer* (2012) <https://assembly.thegef.org/publications/implementing-poznan-strategic-and-long-term-program-technology-transfer>;UNFCCC Technology Executive Committee 'Updated evaluation of the Poznan strategic programme on technology transfer' (18 March 2019) UNFCCC Doc TEC/2019/18/4; UNFCCC Subsidiary Body for Implementation 'Report of the global environment facility on the progress made in carrying out the Poznan strategic programme on technology transfer' (1 May 2015) UNFCCC Doc FCCC/SBI/2015/INF.4.

47 Dussaux et al. (n 31) 4.

48 Seuba counted 141 PTAs with reference to IP protection. Xavier Seuba 'Intellectual property in preferential trade agreements: What treaties, what content?' (2013) 16 *Journal of World Intellectual Property* 240.

49 Seuba (n 49).

protection.[50] An example of a PTA with wide IP coverage is the EU-Japan FTA, which includes a full chapter on IP protection. This chapter widens the IP protection committed to by the parties under TRIPS, mostly with respect to protected geographical indications. While this agreement provides specific treatment for specific industries (again, notably those that require geographical indication protection), the RE sector is entirely ignored.

Another agreement that addresses IP protection very comprehensively is the US-Canada-Mexico (USMCA) FTA. Sixty-four pages long, the USMCA IP chapter covers a wide range of topics. It extends the coverage provided by TRIPS on several issues (e.g. 70 years protection for copyright), and adds up-to-date provisions on current relevant challenges, notably digital trade. Yet, again, there are no specific references to renewable or otherwise green technologies.

A cursory review of other PTAs reveals that the lack of attention to RE-related industries in IP chapters is fairly common. Whether this lack of PCD is problematic or not, however, is debated. On the one hand, if indeed, as indicated by Dussaux et al., strong IP protection is useful for the promotion of RE-related investment, decision-makers will have to start addressing this linkage far more seriously. On the other hand, the lack of clarity regarding the exact impact of IP protection on RE-related investment may justify further research on this linkage, rather than immediate regulation.

Indeed, once (and if) further research provides more conclusive results concerning the positive impact of increased IP protection on RE-related FDI in the Global South, a useful way forward would be to avoid the multilateral route and address IP protection on a bilateral/multilateral basis. Such bilateral/regional talks could isolate IP discussions only to the context of RE-related investment, avoiding the political tensions and suspicions that usually accompany IP-related discussions in multilateral fora.

Moreover, bilateral discussions could lead to a more targeted and tailored outcome, such as assurances made specifically, and exclusively, for pre-approved RE investors/projects that reflect host countries' needs. A better targeted/tailored outcome is necessary also in light of the exact impact that IP protection could play in any given bilateral relationship, which may vary from state to state.[51] This process could be linked to the work of the envisioned committee on EGS, discussed above, which could issue specific IP-related guarantees for any RE investors approved for specific projects.

IV Trade rules on subsidies as international law HCMs

The interaction between trade law on subsidies, on the one hand, and RE policies, on the other, has received significant attention from scholars in recent

50 Seuba (n 49) 245.
51 Dussaux et al. (n 31).

years, including from the author of this study.[52] The debate concerning the impact of subsidies on RE policies[53] can be divided into three broad areas.

First, there are those who ask whether existing trade rules indeed permit the operation of RE support schemes, notably Feed-in Tariffs (FIT) programmes (a question that was by and large settled in the case-law).

Second, there are those who point to the harmful impact of fossil fuels' subsidies on the efforts to increase investment in the RE sector (and on climate change more widely), and advocate for the adoption of new trade rules to address this problem.

Third, there is also the impact that trade remedies may have on the subsidisation of RE equipment exports, notably from China. These three points will be discussed below.

A RE support schemes and the SCM Agreement

The first link between trade rules on subsidies and RE policies has been widely discussed in academic literature.[54] There is no doubt that many states are subsidising the production of RE via support schemes. Whether this form of subsidisation is compatible or not with the instructions of the WTO's Agreement on Subsidies and Countervailing Measures (SCM Agreement) is subject to debate.[55]

In fact, almost every element of the legal definition of the SCM's definition of 'subsidy' is debated in this context. For example, some have claimed that the small size of the RE sector (in comparison with the 'general' energy sector) deems it 'specific' in nature,[56] a determination that may have been

52 Avidan Kent and Vioma Jha 'Keeping up with the changing climate: The WTO's evolutive approach in response to the trade and climate conundrum' (2014) 15(1) *Journal of World Investment and Trade* 245; Avidan Kent 'WTO law on subsidies and climate change: Overcoming the dissonance' (2013) 5 *Trade Law and Development* 344.

53 The reader should be aware that there are different forms of subsidisation for different RE-related activities, from consumption to production, exportation and more. See review in Arunabha Ghosh with Himani Gangania 'Governing clean energy subsidies: What, why, and how legal?' in Gary Hufbauer et al. (eds.) (n 83) 151.

54 Kent and Jha (n 52); Kent (n 52); 3 Luca Rubini 'Ain't wastin' time no more: Subsidies for renewable energy, the SCM Agreement, policy space, and law reform' 15(2) (2012) J. INT'L ECON. LAW 525; Kati Kulovesi 'Real or imagined controversies – A climate law perspective on the growing links between the international trade and climate change regimes' (2014) 6(1) *Trade, Law & Development* 55; Aaron Cosbey and Petros Mavroidis 'A turquoise mess: Green subsidies, blue industrial policy and renewable energy: The case for redrafting the subsidies agreement of the WTO' (2014) 17 *Journal of International Economic Law* 11.

55 Most will agree that it depends on the manner in which subsidies are designs. Some (e.g. Rubini) argue that support schemes could breach the SCM Agreement, while others (e.g. Kulovesi) believe that in practice this is not a real problem.

56 Sadeq Z. Bigdeli 'Incentive schemes to promote renewables and the WTO law of subsidies' in Thomas Cottier et al. (eds.) *International Trade Regulation and the Mitigation of Climate Change* (CUP 2009) 155.

correct when this claim was made (2009) but highly unlikely to hold as the years pass and this sector grows. The SCM's 'benefit' test is also debated. The fact that the purpose of support schemes is indeed to provide an incentive, *beyond market conditions*, is fairly straightforward and hard to deny. On the other hand, Howse (and others) have claimed in the past that the market is already distorted in favour of fossil fuels. Support schemes, therefore, are merely correcting the market, rather than providing a 'benefit' to RE producers.[57]

Whatever authors' views on this matter may be, the WTO's Appellate Body has ruled on this question and stated that support schemes are, in essence, compatible with the SCM Agreement. This is true as long as they are designed fairly and in a non-discriminatory manner (notably through a competitive process (e.g. bidding) and without a local-content requirement).[58] In the context of this study, there is no need to delve any further into this discussion as it is not directly related to international law HCMs. FIT programmes are designed to attract investors, not to incentivise their operation elsewhere. Besides, as stated above, the literature on this topic has vastly grown in the past few years and readers could read more on this topic elsewhere.[59]

B Fossil fuel subsidies and potential reforms to the SCM Agreement

The second issue that concerns the link between RE investment and trade rules on subsidies is related to the damaging role of fossil fuel subsidies. As discussed elsewhere in this study, fossil fuel subsidies are forming a barrier for investment in RE, notably as they reduce the price of fossil fuels and affect the competitive position of RE projects. States wishing to support their home RE investors will therefore do wisely to negotiate the removal of fossil fuel subsidies in host countries. Rules on this matter could thus be regarded as HCMs.

The removal of fossil fuel subsidies is a hotly debated topic in global trade negotiations. A group of 12 WTO Member States (including developing, Sub-Saharan states[60]) issued in 2017 a joint communique calling for a reform to the regulation of fossil fuel subsidies.[61] This communique is very general, vague and limited in content, essentially calling for the elimination

57 Robert Howse, *Climate Mitigation Subsidies and the WTO Legal Framework: A Policy Analysis* (IISD 2010) <https://www.iisd.org/pdf/2009/bali_2_copenhagen_subsidies_legal.pdf> 5.

58 *Canada—Certain Measures Affecting the Renewable Energy Generation Sector* (6 May 2013) WT/DS412/AB/R and WT/DS426/AB/R, Report of the Appellate Body.

59 See sources in footnotes 52 and 54.

60 This call was made by an informal group of states, self-referred to as 'Friends of Fossil Fuel Subsidy Reform'. This group includes the following states: Costa Rica, Denmark, Ethiopia, Finland, New Zealand, Norway, Sweden, Switzerland and Uruguay.

61 The communique is found here: <http://fffsr.org/communique/> 'fossil fuel subsidies reform ministerial statement' (2017) WTO Doc WT/MIN(17)/54.

of fossil fuel subsidies. As James Bacchus explains, the negotiations for these new rules are expected to be far more complicated than this communique implies.[62] Notably, questions such as what kind of subsidies will be included under this reform, or how these subsidies will be measured (e.g. should externalities be included or not?) will have to be determined.[63]

Bacchus himself discusses some of the options that are available under the trading regime.[64] First, there is the option of using existing WTO rules on subsidies as a basis to challenge the legality of fossil fuel subsidies. Joseph Stiglitz famously claimed in the past that US firms that rely on cheap fossil fuels are effectively being subsidised, as they are not required to pay for the environmental consequences of their operation. He therefore claims that based on WTO Law, other states should impose a 'tax' on these US firms in order to offset this 'subsidy'.[65] Somewhat surprisingly,[66] in its last days in power (and after losing the election), the US Trump Administration submitted a somewhat similar proposal to the WTO.[67]

62 James Bacchus 'WTO negotiations on fossil fuels subsidy must begin now' (2018) *ICTSD* <https://www.ictsd.org/opinion/wto-negotiations-on-fossil-fuel-subsidy-reform-must-begin-now>.
63 Bacchus (n 62).
64 Bacchus (n 62).
65 Joseph Stiglitz 'A new agenda for global warming' (2006) *Economists' Voice* <http://carbon-price.com/wp-content/uploads/2006-07-stiglitz-a-new-agenda-for-global-warming.pdf> 2.
66 This proposal is surprising, given the US' own emissions, which would suggest that the US could be targeted for breaching the SCM Agreement.
67 *Advancing Sustainability Goals Through Trade Rules to Level the Playing Field: Draft Ministerial Decision* (17 December 2020) WT/GC/W/814 <https://docs.wto.org/dol2fe/Pages/SS/directdoc.aspx?filename=q:/WT/GC/W814.pdf&Open=True>.

> Recognising that the value of a subsidy is not the cost of the subsidy to the granting government but the benefit received by an industrial entity;
>
> Realising that industries located in certain countries benefit from weak or unenforced environmental laws and regulations by not being required to incur, and properly internalize, the costs of preventing or remediating environmental damage resulting from their production processes;
>
> Recalling that the Agreement on Subsidies and Countervailing Measures (ASCM) permits Members to impose a countervailing duty to offset a government subsidy benefit received by an exporter to ensure that domestic producers should only have to compete with those other firms that are subject to the same competitive market conditions;
>
> Accepting that imposing countervailing measures on subsidies that take the form of weak or unenforced environmental standards would promote stronger environmental standards and enforcement, would encourage the proper internalization of environmental costs into the calculations of production costs and would correct policies that create transaction-specific market inefficiencies which thereby distort trade;
>
> Decides as follows:
>
> 1. The failure of a government to adopt, maintain, implement and effectively enforce laws and regulations that ensure environmental protections at or above a threshold of fundamental standards shall constitute an actionable subsidy under the ASCM.

These claims and proposals are without a doubt appealing. It remains questionable however how effective current rules are in this respect, or whether these claims have any legal basis at all. To begin with, claiming that such alleged 'subsidies' are specific in nature is unconvincing. The SCM Agreement applies only to subsidies that are specific to a certain industry or region. Fossil fuel 'subsidies', on the other hand, are widely available to all energy consumers. As Marc Benitah notes, fossil fuel 'subsidies' are also unlikely to satisfy the 'financial contribution' test, which does not apply to a situation in which a state simply refuses to tax a certain activity.[68]

Another option presented by Bacchus is simply to reform WTO Law. For example, Bacchus suggests 'waiving' certain requirements that currently prevent a potential challenge to the legality of fossil fuel subsidies under the SCM Agreement (e.g. the 'specificity' and the 'adverse affect' requirements).[69] Another possibility mentioned by Bacchus is simply to ban such subsidies, or, as proposed by Trachtman, to design altogether new rules for energy subsidies.[70] Indeed, several authors have made more detailed proposals in this respect.[71] These options are certainly appealing, even if politically unfeasible in light of existing WTO (and UNFCCC) geo-political divisions.

To conclude, there is no doubt that states will have to address the topic of fossil fuel subsidies in a direct and effective manner, whether through the WTO (as suggested by Bacchus) or, more realistically, via bilateral/regional trade deals. The regional/bilateral route seems far less politically complex. Home (Western) countries could agree with their trading parties on the mutual elimination of fossil fuel subsidies, in exchange for 'carrots' in the shape of opening their own lucrative markets. If the global energy market is indeed turning 'green', and if it is as lucrative as it appears to be, the elimination of barriers such as fossil fuel subsidies will compensate home countries for whatever trade concessions they will have to make in order to achieve this

2. If an industry disproportionately benefits from pollution controls or other environmental measures set below a threshold of fundamental standards, a Member may impose a countervailing duty equal to the benefit received by the industry when the goods from such an industry enter the Member's customs territory.

68 Marc Benitah 'Climate change: Is there place for a WTO Anti-Subsidy strategy? (29 December 2006) *Global Subsidies Initiative* <https://iisd.org/gsi/commentary/climate-change-there-place-wto-anti-subsidy-strategy>.

69 Bacchus (n 62).

70 Joel P. Trachtman 'Fossil fuel subsidies reduction and the World Trade Organization' (ICTSD 2017) <https://www.ictsd.org/themes/climate-and-energy/research/fossil-fuel-subsidies-reduction-and-the-world-trade-organization>; See also Gary Horlick 'The WTO subsidies agreement can be changed to discipline fossil fuel subsidies' (ICTSD, 22 August 2017) <https://www.ictsd.org/opinion/the-wto-subsidies-agreement-can-be-changed-to-discipline-fossil-fuel-subsidies>; Heloisa Pereira 'How the WTO can help tackle climate change through fossil fuel subsidy reform: Lessons from the fisheries negotiations' (ICTSD 2017) <https://www.ictsd.org/themes/climate-and-energy/research/how-the-wto-can-help-tackle-climate-change-through-fossil-fuel>.

71 Ibid.

goal. Importantly, the 'bilateral' elimination of fossil fuels' subsidies will in fact be multilateral in nature, as countries cannot eliminate such subsidies for the benefit of one country alone.

C *Export subsidies*

Recent years have seen a significant, welcome drop in the price of RE equipment. This drop is important as it facilitates the dissemination of RE across the world (including to poorer parts of the world) and leads to an improved competitive position for RE technologies. The drop in prices is mostly, but not only, occurring due to technological improvements. The fact that states – notably China – have been providing financial support for RE equipment producers and exporters has also been contributing to this trend.

Naturally, the support provided by exporting states is damaging to the commercial interests of producers from other countries. RE equipment producers from developed countries have therefore lobbied their governments to put a stop to China's subsidisation. The result of this effort was an all-in trade war, notably the imposition of anti-dumping and countervailing duties on imported RE equipment, which was followed by trade retaliation measures adopted by China.

It is questionable whether this result is useful for the global promotion and dissemination of RE. Some will claim that ensuring a level playing field is necessary for the long-term competitiveness of the free market and its sustainability. Others (including the author of this study) have argued that the urgency involved in climate change requires urgent intervention, and waiting for the (long-term) impact of the free market's invisible hand might be disastrous for the environment. As noted by Cottier[72]:

> [w]hile these measures can be understood from the point of view of protection of domestic labour and industry, they increase costs and make renewable energy equipment more expensive to investors and consumers. From the point of view of climate change CO_2 emissions reduction, measures taken thus run counter to the goals of decarbonisation and the dissemination of renewable energy production. [...] From the point of view of environmental law, easy recourse to anti-dumping and countervailing duties is counterproductive.

A similar point was also made by the author of this study in a previous paper.[73] There is no doubt that the reduction to the price of RE-related equipment through export subsidies could encourage RE-related FDI.

72 Thomas Cottier 'Renewable energy and WTO Law: More policy space or enhanced disciplines? (2014) 1 *RELP* 40, 43.

73 Kent (n 52) 378.

This is true for the same reason that low tariffs on EGS, or any other form of financial support, are reducing the cost of investment. It is clear that such measures are damaging in the longer term due to their negative impact on competition. The context of RE, however, is different. The problem at hand is urgent and requires a fast and substantial dissemination of RE technologies,[74] and trade remedies that are imposed on the subsidisation of EGS production are inhibiting this process. Indeed, representatives from the Solar Trade Association commented with respect to the EU's decision to extend the use of countervailing and anti-dumping duties on Chinese solar panels[75]:

> *These price controls on imports of Chinese solar panels need to be dropped. Europe is currently paying far more than it should for its solar – and that applies both to our homeowners and our governments. [...]*

The subsidisation of the RE industry and its compatibility with WTO Law is discussed to a great extent in the literature.[76] Due to space limitations, the author will not expand further on this topic. The reader nevertheless should be aware that these issues are relevant for the regulation of RE, and that they do affect investment in this field.

V Trade in services

Trade in environmental services is another key issue that will affect the facilitation of investment in the RE sector. The rules on trade in services are

74 Avidan Kent and Vioma Jha 'Keeping up with the changing climate: The WTO's evolutive approach in response to the trade and climate conundrum' (2014) 15(1) *Journal of World Investment and Trade* 245; Avidan Kent 'WTO law on subsidies and climate change: Overcoming the dissonance' (2013) 5 *Trade Law and Development* 344. See also Thomas Brewer as well argues that 'Market forces alone are insufficient to bring about a rapid deployment of sustainable energy technologies' in Thomas Brewer 'International technology diffusion in a sustainable energy trade agreement: September 2012' in Gary Hufbauer et al. (eds) (n 83) 369.

75 Solar Trade Association, *European Commission Extends 'Unfair' Import Tariff on Chinese Solar Panels* (Solar Trade Association, 7 December 2015) <http://www.solar-trade.org.uk/european-commie-solar-panels/>; see also views of STA's representatives in Sonia Dunlop 'Why EU tariffs on solar panels need to end' (The New Economy, 8 February2016)<http://www.theneweconomy.com/energy/eu-import-tariffs-on-chinese-solar-panels-need-to-end>.

76 See for example: Aaron Cosbey and Petros Mavroidis 'A turquoise mess: Green subsidies, blue industrial policy and renewable energy: The case for redrafting the subsidies agreement of the WTO' (2014) 17(1) *Journal of International Economic Law* 11; Luca Rubini 'Ain't Wastin' time no more: Subsidies for renewable energy, the SCM agreement, policy space, and law reform' (2012) 15(2) *Journal of International Economic Law* 525; Sadeq Bigdeli 'Clash of rationalities: Revisiting the trade and environment debate in light of WTO disputes over Green Industrial Policy' (2014) VI(1) *Trade Law and Development* 177.

essentially referring to the right of a service provided from state A, to provide a service in state B, whether in the form of:

> (1) Cross-border provision of services, i.e. the service provider and the consumer do not physically cross a border (Mode 1),
> (2) Consumption abroad, i.e. the consumer is travelling to the service provider's state (Mode 2),
> (3) Commercial presence, i.e. the provider is establishing itself in the territory of the consumer's state (Mode 3), or
> (4) Presence of natural persons, i.e. the provider is travelling to the consumer's state for the sake of providing the service (Mode 4).

These rules, therefore, address a wide variety of issues ranging from visa requirements, licenses and permits to operate, recognition of qualifications, and more. These rules are clearly related to FDI, especially where the investor (e.g. mother company) is maintaining close working links with the 'investment' (e.g. the energy facility established on the ground).

A Trade in environmental services: benefits for the environment, and beyond

As products are becoming increasingly technologically intensive (as is the case in the energy sector), it is becoming clearer that the provision of accompanying services can no longer be regarded separately from the product itself. Some authors have referred to this phenomenon as the 'servicification' of goods; it is claimed that services now account for a quarter of goods' value.[77] In the context of environmental protection, the added value of services includes *inter alia* benefits such as the transfer of green technologies, the reduction of emissions,[78] improving efficiency and competition in importing markets, greater availability of infrastructure, support for local industry, fostering green innovation and more.[79] In the wider context of development the value of services was described in the following words[80]:

> Attaining the SDGs is to a significant extent a services agenda. Eliminating poverty and hunger, improving health and educational outcomes,

77 Michael Alder et al. 'Climate change and services trade: What role for the GATS?' in Panagiotis Delimatsis (ed.) *Research handbook on climate change and international trade* (EE 2016) 182–183.
78 Alder et al (n 77) 182–183.
79 Nora Dihel 'Understanding trade in environmental services: Key issues and prospects' in Olivier Cattaneo et al. (eds.) *International Trade in Services: New Trends and Opportunities for Developing Countries* (World Bank 2010) 319–321.
80 Matteo Fiorini and Bernard Hoekman, *Services Trade Policy and Sustainable Development* (2018) <http://respect.eui.eu/wp-content/uploads/sites/6/2019/02/Fiorini_Hoekman_Services_SDGs_2018.pdf> 1–2.

> or reducing regional inequalities will require boosting services capacity and the productivity of a range of services activities, including transport, distribution, logistics, ICT, vocational training, medical services and so forth.

The need to address trade in environmental services in any future green agreement is evident. According to a certain study, some services are 'indispensable for the trade in environmental goods [...] Or to put it succinctly: without services, no trade'.[81] This same study identified as 'indispensable' the following services: 'assembly and installation, technical testing and analysis, educational services, advisory and consultative services, managing and repairs, computer services, R&D, and "traditional" environmental services'.[82] More specifically with respect to RE projects, Monkelbaan lists a variety of relevant services, including assessment of wind and solar resources, site analysis, project development and financing, construction, installation and maintenance, quality and performance testing, as well as a line of legal, engineering and technical services, and more.[83]

According to a United States International Trade Commission study, most RE service providers originate from either the US or the EU[84]; a fact that implies the need to liberalise this type of trade and also facilitate the reach of these services to the Global South. As stated in an OECD report, the benefit of trade in environmental services in the context of development is evident as 'firms exporting environmental consulting and engineering services [tend, AK] to be larger, more productive, and to pay higher salaries than their domestically focussed counterparts'.[85]

B The WTO and the trade in environmental services

An OECD report on the trade in environmental services recommends a quicker reduction to barriers to the trade in environmental services.[86] A review of certain key states' GATS commitments conducted by Monkelbaan

81 Camilla Prawitz et al., *Making Green Trade Happen: Environmental Goods and Indispensable Services* (Swedish National Board of Trade 2014) 2.

82 Prawitz et al. (n 81) 2.

83 Joachim Monkelbaan 'Trade in sustainable energy services' in Gary Hufbauer et al. (eds.) *The law and economics of a sustainable energy trade agreement* (CUP 2013) in Box 3.1, and p. 95. See also Lisa Ferens Alejandro et al., *Renewable Energy and Related Services: Recent Developments* (USITC 2013) xv.

84 Alejandro et al. ibid. Since the USITC's research was published, China has also emerged as a major provider or services in this field.

85 OECD Joint Working Party on Trade and Environment 'Trade in services related to the environment' (2017) COM/TAD/ENV/JWPTE(2015)61/FINAL <http://www.oecd.org/officialdocuments/publicdisplaydocumentpdf/?cote=COM/TAD/ENV/JWPTE(2015)61/FINAL&docLanguage=En> 30.

86 OECD (n 85) 30.

confirms that most states have not made significant reductions to barriers vis-à-vis the main modes of services.[87]

The reasons that barriers to trade in environmental services exist vary. Notably, many argue that the classification list of services relied on by the WTO Member States (W/120) is dated and does not reflect modern environmental services.[88] Indeed, while certain general classifications capture services that are relevant for the RE industry, their lack of specificity leads to the 'dual-use' problem, i.e. such classifications cover both environmental and non-environmental services, making environmentally targeted concessions harder to achieve.[89] The 'environmental services' section in the W/120 list also seems dated and includes only four categories: 'sewage', 'sanitation', 'refuse disposal' and 'others'.[90]

More specific attempts to address this problem were made under the WTO within the institution's Committee on Trade and Services Special Session (CTS SS). An EU Communication[91] has called for the revision of the W/120 classification list and the adoption of 'a new classification for "core" environmental services which better reflects the trade and sectoral realities'.[92] Another Communication from a group of developed states[93] (the EU, Taiwan, the US and Japan) also specifically addressed 'environmental infrastructure services', although no specific reference was made to RE projects. Other proposals in this respect were also made by others, notably Switzerland and Canada.[94] At the time of writing, discussions under the CTS SS are no longer focused on this topic, and WTO negotiations (as discussed above) are mostly focused on the liberalisation of trade in goods.[95]

87 Monkelbaan (n 83) 101.
88 Joachim Monkelbaan, *Trade in Sustainable Energy Services* (ICTSD 2013) 7–8; Alder et al. (n 77) 184.
89 Monkelbaan (n 88) 8.
90 See sectors 6(A), (B) (C) and (D). As explained by Adler et al. (n 77), the 'others' category refers to services related to the cleaning of exhaust gases, noise abatement, nature and landscaping protection, and 'other environmental protection services not elsewhere classified'. As noted by these authors, there is indeed one reference to climate – under 'nature and landscape protection' services, services that are related to the research of climate are covered. These authors are noting that this mentioning is not visible in the W/120 list itself. Adler et al. (n 77) 184.
91 Communication from the EC: GATS 2000: Environmental Services, WTO Doc S/CSS/W/38 (22 December 2000).
92 Communication from the EC (n 91).
93 Communication from Australia, The European Communities, Japan, New Zealand, The Separate Customs Territory of Taiwan, Penghu, Kinmen and Matsu and The United States, WTO Doc TN/S/W/28 (11 February 2005).
94 See in more detail in Adler at al. (n 77) 185.
95 OECD 'Opening Markets for Environmental Goods and Services' (2017) OECD Policy Brief <http://www.oas.org/dsd/Tool-kit/Documentos/ModuleIIIdoc/Opening%20Markets%20for%20Environmental%20Goods%20and%20Services.pdf> 5.

One ray of light in this rather bleak reality could be the ongoing negotiations of a Plurilateral Trade in Services Agreement (TiSA). The TiSA negotiations currently only include 23 states. While these states (including trade giants such as the EU, the US and China) represent 70% of the world's trade in services, they are mostly services exporter states, rather than importers (although some 'importers' are included), and the Global South's participation is fairly limited[96] (e.g. Mauritius being the only African country currently participating in these negotiations). According to several sources, the EU (and possibly other states as well[97]) is pushing for the inclusion of a sectoral annex on trade in environmental services.[98] The adoption of an up-to-date, specific annex on environmental services will certainly be useful for RE investment, should it include relevant concessions. However, whether this development will take place remains to be seen.

C Trade in environmental services in other trade agreements

A cursory review (based on the WTO's search engine) of environmental services in Regional Trade Agreements reveals that, where included in states' concessions, they often reflect the same W/120 classification and include the 'usual' areas such as water treatment, sanitation and sewage. According to these results, there are no specific references to RE. An OECD report explains this focus as a result of a dated approach and the following of 'old habits'[99]:

> Trade negotiators have long taken environmental services to refer to the few sectors described under division 94 of the UN's Central Product Classification (CPC), namely activities in relation to wastewater treatment, waste collection and management, remediation, sanitation and similar services, and a few other environmental-protection-related services.

On some occasions, it seems that the Member States recognise the important role of trade in environmental services but are unwilling to take the next step. Chapter 24 of the USMCA states that[100]:

96 Notable examples are Pakistan, Colombia, Panama and Mauritius.

97 According to this source, Canada as well is supporting the liberalisation of trade in environmental services, Elina Viilup, The Trade in Services Agreement (TISA): An end to negotiations in sight? (2015) Policy Department, Directorate-General for External Policies <http://www.europarl.europa.eu/RegData/etudes/IDAN/2015/570448/EXPO_IDA%282015%29570448_EN.pdf> 18.

98 Viilup (n 97) 15, 18; Rachel F. Fefer 'Trade in Services Agreement (TiSA) negotiations: Overview and issues for congress' (2017) US Congressional Research Service 7–5700 <https://fas.org/sgp/crs/misc/R44354.pdf> 7.

99 OECD (n 85) 3.

100 Emphasis is not in the original. See Art 24.24(1) and (2) of the USMCA.

> The Parties recognize the importance of trade and investment in environmental goods and services, *including clean technologies*, as a means of improving environmental and economic performance, contributing to green growth and jobs, and encouraging sustainable development, while addressing global environmental challenges [...] Accordingly, the Parties shall strive to facilitate and promote trade and investment in environmental goods and services.

On a more optimistic note, Art 24.24(3) USMCA states that the USMCA's Environment Committee 'shall consider issues identified by a Party' in this context, including non-tariff trade barriers.[101] Art 24.24(4) further adds that 'Parties shall cooperate in international fora on ways to further facilitate and liberalize global trade in environmental goods and services, and may develop cooperative projects on environmental goods and services to address current and future global environmental challenges'.[102]

Similarly soft language is found also in the EU's trade agreements. For example, Art 16.5(c) of the EU-Japan FTA states that the parties

> shall strive to facilitate trade and investment in goods and services *of particular relevance to climate change mitigation*, such as those related to *sustainable renewable energy and energy efficient goods and services*, in a manner consistent with this Agreement.

Other examples of similarly soft, non-committed language are also found in trade agreements concluded by other countries.[103] A more ambitious attempt to liberalise trade in EGS is found in the New Zealand-Taiwan FTA. This FTA provides a specific list of goods (see Annex 7 of this Agreement), and with respect to services it states that this agreement shall 'facilitate the movement of business persons involved in the sale, delivery or installation of environmental goods or the supply of environmental services in accordance with Chapter 14 (Temporary Entry of Business Persons)'.[104] The term 'Environmental Services' is defined in a footnote to this chapter as[105]:

> environmental services refers to *services directly related to the investment, sale, delivery or installation of environmental goods defined in subparagraph 2(a)* of this Article, as well as services that fall under the WTO classification MTN.GNS/w/120 or the Provisional Central Product Classification, 1991

101 Art 24.24.(3) USMCA.
102 Art 24.24.(4) USMCA.
103 See for example Art 17.4 of the Canada-Korea FTA.
104 Art 3(2)(b), Chapter 17 New-Zealand-Taiwan FTA.
105 Footnote 14, Art 3(2)(b), Chapter 17 New-Zealand-Taiwan FTA.

While the latter part of this definition refers to the W/12O list and does not add much to the purpose of this study, the first part is nevertheless interesting. The definition referred to in subparagraph 2(a) for the term environmental goods (defined also in a footnote) is 'those which positively contribute to the green growth and sustainable development objectives of the Parties'.[106] This chapter therefore presents a very wide definition of environmental services, one that goes far beyond the usual W/120 list and seems much more appropriate for the liberalisation of EGS. The fact that this agreement provides a clear list of environmental goods, as well as the reference to Chapter 14 of this agreement, is helpful as it provides clarity and some level of concreteness to these commitments.

VI RE regulation in FTAs

The previous parts address specific trade-related issues that are of direct relevance to the promotion of RE-related investment in the developing world. The following section will complete this discussion, by attempting to scope more widely for examples of language addressing RE-related trade and investment in FTAs.

There are very few attempts to regulate RE-related trade and investment through FTAs, and even fewer attempts to promote any form of HCMs. This is not to argue that FTAs are entirely silent on issues such as climate change, and indeed, authors have shown that many FTAs do address climate change in one way or another.[107] While the manner in which these FTAs address climate change differs, the review below demonstrates that the overall level of commitment demonstrated in these treaties has so far been very weak.

To begin with, many treaties refer to broad terms such as 'environmental protection' and 'sustainable development' as treaty objectives. For example, one of the objectives of the USMCA is to[108]:

> PROMOTE high levels of environmental protection, including through effective enforcement by each Party of its environmental laws, as well as through enhanced environmental cooperation, and further the aims of sustainable development, including through mutually supportive trade and environmental policies and practices.

The European Free Trade Association (EFTA)-Philippines FTA equally states that the parties are 'AIMING to create new employment opportunities, improve living standards, and raise levels of protection of health and

106 Footnote to Art 3(2)(a), Chapter 17 New Zealand-Taiwan FTA.
107 Markus Gehring et al., *Climate Change and Sustainable Development Measures in Regional Trade Agreements* (ICTSD 2013).
108 USMCA, preamble.

safety, and of the environment'.[109] Other treaties, notably those signed by the EU, also routinely refer to such terms as 'sustainable development' as a leading principle/objective of their FTAs.[110]

Other attempts to address the link between trade and the environment include the adoption of exceptions for measures that are related to climate change (notable via Art XX General Agreement on Tariffs and Trade (GATT)-like exceptions),[111] or through clarifications concerning the importance of complying with environmental agreements.[112] These measures, it should be stated, are 'defensive' in nature as they are aimed at ensuring states' right to regulate, and also, notably, that their climate policies will not conflict with their trade law obligations.[113] As such, these provisions can hardly be considered as HCMs, as they are not aimed at investment promotion (at least not directly). These safeguards are nevertheless important: they are crucial for the implementation of the notion of PCD as they protect a wide range of interests.

More specifically, with respect to the RE sector, only very few FTAs explicitly and directly address this sector. Most of these examples are somewhat weak in nature. The author of this study describes elsewhere these attempts as 'mostly declaratory, vague, and non-committed, and [...] often built on future arrangements which may or may not be negotiated. As such their usefulness could be doubted'.[114] One such example is Art 275(4) of the EU-Colombia-Ecuador-Peru FTA[115]:

> Considering the global objective of a rapid transition to low-carbon economies, the Parties will promote the sustainable use of natural resources and will promote trade and investment measures that promote and facilitate access, dissemination and use of best available

109 EFTA-Philippines FTA, preamble.

110 See for example the EU-Chile Association Agreement and the CARIFORUM-EC Economic Partnership Agreement, both are referring to 'sustainable development' as a leading principle. Chile-EC Association Agreement, Preamble, Art 1(2); CARIFORUM-EC Economic Partnership Agreement, Art 1 and 3.

111 E.g. EFTA-Philippines FTA, Art 2.17; Colombia-Ecuador-EU-Peru Trade Agreement, Art 106.

112 See for example: Central American Common Market (CACM)-EU Association Agreement, Art 287.

113 The reader should note that existing exceptions may not always be useful in avoiding conflicts between trade obligations and climate change obligations, and much depends on how such provisions are to be interpreted, the manner in which preambles' statements and treaty objectives are understood by panels, or even whether existing legal exceptions apply (e.g. most will agree that GATT XX exceptions do not cover the SCM agreement).

114 Avidan Kent 'Policy coherence and the promotion of foreign direct investment in the renewable energy sector: Lessons from Europe' (2018) The Yearbook on International Investment Law and Policy 409.

115 See also: Columbia-Ecuador-EU-Peru Trade Agreement (n 120), Art 24.12; EU-Canada Comprehensive Economic and Trade Agreement (CETA), Art 24.9.

> technologies for clean energy production and use, and for mitigation of and adaptation to climate change.

Another interesting example is Article 138 of the CARIFORUM-EC agreement ('Cooperation on eco-innovation and renewable energy'), which directly addresses RE. The language of this Article is nevertheless non-committed and relies on future arrangements ('the Parties agree to cooperate, including by facilitating support, in the following areas [...] projects related to energy efficiency and renewable energy'[116]). Elsewhere, the CARIFORUM-EC agreement adds a further relevant (yet non-committed) reference to trade in EGS (including specifically to RE-related EGS)[117]:

> The Parties and the Signatory CARIFORUM States are resolved to make efforts to facilitate trade in goods and services which the Parties consider to be beneficial to the environment. Such products may include environmental technologies, renewable- and energy-efficient goods and services and eco-labelled goods.

Similar examples are mostly found in other agreements concluded between the EU and other parties (e.g. Art 288 of the Central American Common Market (CACM)-EU Association Agreement,[118] or Art 24.9 of the EU-Canada FTA (CETA)), but not only (e.g. Art 20.15 of the Trans-Pacific Partnership, and Article 24.24 in the USMCA).

All of these examples specifically refer to the increased dissemination of RE through liberalised international trade,[119] but nevertheless do not provide any concrete rules on this matter. If to classify these efforts under UNCTAD's typology, it can be said that these efforts fall under the first type of HCMs, i.e. 'vague, non-committed and declaratory language, which at least for now does not mean much in practice'.[120]

More recently (2018–2019), the EU has started to adopt a new and more ambitious approach. The EU's FTAs with Singapore (2018) and Vietnam (2019) both include comprehensive chapters on non-tariff barriers to trade and investment in RE generation. The declared objective of these chapters is in line with the notion of PCD, addressing environmental and developmental concerns. These chapters include several important provisions, such as a prohibition on Local Content Requirements (LCR), non-discrimination

116 CARIFORUM-EC Agreement (n 119), Art 138(b).

117 Ibid., Art 183(5).

118 Ibid., Art 288(c)-(d).

119 The recent USMCA takes a similar stand but refrains from addressing climate change (probably for political reasons related to the stand of the US on this topic). See Art 24.24 USMCA.

120 Kent (n 114).

and transparency vis-à-vis authorisation, certification and licensing, as well as technical standards.

The adoption of a dedicated, comprehensive chapter on RE is certainly a step in the right direction. It is nevertheless important to state that some of these obligations are at least partly covered by other international agreements. For example, reliance on international standards is required by the WTO TBT Agreement. Similarly, LCR is addressed by agreements such as the TRIMS and the SCM, and non-discriminatory treatment is also widely banned under a variety of WTO rules. Despite these issues, certain authors have viewed this as a positive development, stating that 'it firmly commit[s] to the effort of eliminating greenhouse emissions [...and] approaches (clean energy from a more integrated trade and investment perspective)'.[121] In other words, even if not ambitious in content, the PCD-led approach is noticeable, a fact that on its own represents an improvement.

VII Trade law and Article 6 of the Paris Agreement

Another potentially relevant trade-related mechanism is found outside of international trade law: Article 6 of the Paris Agreement establishes a Sustainable Development Mechanism (SDM). The specific shape and rules of the SDM have not yet been agreed, but in essence it is expected to succeed the Kyoto Protocol and fulfil a similar goal: promote clean investment, and allow states more flexibility in meeting their self-determined emission reduction goals.[122] Art 6.4 of the Paris Agreement is especially relevant to this study: it is expected to follow the footsteps of Kyoto's Clean Development Mechanism (CDM) and promote investment in RE in the Global South.

Importantly, the SDM will rely on carbon trading as a way of incentivising foreign investment. In essence, the reduced emissions achieved through RE investments are quantified and packaged as emission reduction units, which are then sold in carbon markets, topping up investors' returns on their investment.

International trade law can play an important role in facilitating the operation of the SDM and improving its efficiency. Cosbey and Marcu have pointed to the fact that traded emission units cannot be defined as 'goods' and therefore may not be covered by the GATT.[123] The GATT's protection (e.g. from non-discriminatory treatment, or quotas), therefore, may not be

121 Anna-Alexandra Marhold 'Externalising Europe's energy policy in EU Free Trade Agreements: A cognitive dissonance between promoting sustainable development and ensuring security of supply?' (2019) Economic and Non-economic Values and Objectives in the EU's International Trade: Actors and Processes in Resolving Normative Tensions 12.

122 See more on the SDM in the chapter on environmental law, in this study.

123 Aaron Cosbey and Andrei Marcu 'The Paris Agreement's article 6 and the WTO: Points of convergence' (2020) CIGI <https://www.cigionline.org/articles/paris-agreements-article-6-and-wto-points-convergence>.

of use. These authors propose that the WTO Member States adopt a formal understanding, confirming that carbon units are 'goods', thus bringing emission units under the scope of the GATT. Alternatively (and possibly more realistically), the UNFCCC Member States could incorporate some of the more relevant GATT disciplines into the specific rules that will govern the SDM. As noted by certain authors, special attention should be given to the role of discriminatory policies, which might be justified in some cases (notably 'hot air' credits, or units based on reductions of greenhouse gases (GHG) that are flooding the market).[124]

VIII Trade law, renewables and policy coherence for development

The above review is not a comprehensive overview of the interaction between trade law and investment in the RE sector. Such a review is beyond the scope of this study, and some trade-related issues without a doubt have been overlooked.[125] The review, however, is sufficient to demonstrate that while trade treaties could be relied on as HCMs in many different ways, in practice they fail to deliver. It is true that there are certain areas in which more research is necessary before any attempts are made to operationalise these tools and apply them in treaties (notably IP protection). Many of these tools are not yet tested, and the impact is not yet clear. At the same time, it is clear that international trade is one of the many facets that could be utilised for the promotion of RE-related FDI and more attention should be given to the linkage between trade and investment in the RE sector. The fact that so little research has been done so far on the interaction between Art 6 of the Paris Agreement and international trade law is therefore worrying.

Reasons for the failure of states to address the link between trade and the environment range from 'traditional' North-South tensions and mistrusts, to competing states' interests and even conflicting interests within states.[126] But, regardless of these reasons, the unfortunate result is one; the potential embedded in trade law as an RE supporting HCM is far from fulfilled.

124 Cosbey and Marcu (n 123); Ilmi Granoff 'Trade implications of climate policy after the Paris outcome' (2016) 129 *Trade Hot Topics* <https://www.oecd-ilibrary.org/docserver/5jlwz7lt4k9q-en.pdf?expires=1591261684&id=id&accname=guest&checksum=7D74580D7D05A688A99FE55D8DBD8F8E> 5.

125 For a more comprehensive review see Susanne Dröge, Harro van Asselt, Kasturi Das and Michael Mehling 'Mobilising trade policy for climate action under the Paris agreement options for the European Union' (2018) SWP Research Paper <https://www.swp-berlin.org/fileadmin/contents/products/research_papers/2018RP01_dge_etal.pdf>; Clara Brandi, trade 'elements in countries' climate contributions under the Paris agreement' (2017) ICTSD Climate and Energy <https://euagenda.eu/upload/publications/untitled-81229-ea.pdf>.

126 For example, Marhold discusses the tensions between the EU's environmental aspirations, on the one hand, and its desire to secure energy sources, on the other (Marhold (n 121)).

The manner in which trade law currently addresses this objective is mostly vague, declaratory and ineffective. The novel approach attempted by the EU is encouraging to a certain extent as it demonstrates both movement in the right direction and an appetite to explore this avenue. In terms of results, however, even the new EU agreements with Singapore and Vietnam are hardly an achievement.

The call for increased cooperation between the trade and climate legal regimes is not new. Brandi calls for an increased institutional interaction between the WTO and the UNFCCC[127]:

> This could involve a stronger focus on climate change in the WTO's Committee on Trade and Environment. Improved coordination and exchange of information on NDC implementation between existing bodies in the trade regime and the climate regime would also be a positive help, for instance the WTO's Committee on Trade and Environment or the Trade Policy Review Mechanism, on the one hand, and the UNFCCC's Subsidiary Body on Scientific and Technological Advice, specifically the forum on the impact of the implementation of response measures, on the other hand.

Others have called for more ambitious regulatory reform, in the shape of a new international agreement: the Sustainable Energy Trade Agreement (SETA).[128] This initiative proposes reform of legal areas such as trade in services, subsidies, government procurement and technical regulation. It further proposes that the SETA could include important HCMs that are relevant for effective PCD, notably in the form of technical assistance, capacity building, bilateral international technology cooperation arrangements and development financing programmes.[129]

There is little doubt that such a reform would be a good step forward: it presents a thorough review of the interaction between trade and RE. The SETA proposal, however, can only be viewed as a point of departure. As reviewed in this study, the interactions that trade law must engage with are far wider, and include *inter alia* areas such as investment, insurance, development, human rights and environmental law. As wide as it is already, in order to be truly effective and representative of the notion of PCD, the future regulation of RE-related investment will have to address these issues as well.

127 Brandi (n 24) 28. The author of this study have made similar recommendation in his PhD dissertation 'International trade, investment, and climate change: A tale of legal and institutional fragmentation' (submitted at the University of Cambridge 2014).

128 Gary Hufbauer, Ricardo Melendez-Ortiz and Richard Samans (eds.) *The Law and Economics of a Sustainable Energy Trade Agreement* (CUP 2016).

129 Brewer (n 74) 372.

6 International development law

Tackling trade-offs through policy coherence

I Introduction

Another subject-area that is of great importance to this study is the laws and policies that are related to international development. Conveniently, the term Policy Coherence for Development (PCD) – which is the topic of this study – was developed within the context of international development law, and was aimed first and foremost at improving the effectiveness of this field of law. Notably, significant efforts have been made at improving the effectiveness of international assistance[1] and increasing policy coherence in the context of development finance. The Monterrey Consensus of the International Conference on Financing for Development (2002) addresses the significance of coherence on the international level[2]:

> [The signatory states shall] underline the importance of continuing *to improve global economic governance* and to strengthen the United Nations leadership role in promoting development. With the same purpose, efforts should be strengthened at the national level to *enhance coordination* among all relevant ministries and institutions. Similarly, *we should encourage policy and programme coordination of international institutions and coherence* at the operational and international levels to meet the Millennium Declaration development goals of sustained economic growth, poverty eradication and sustainable development.

The Monterrey Consensus further highlights the important role that trade, investment, finance and development institutions should all play in this respect, *inter alia*, as facilitators of activities such as international trade and investment. The Monterey Consensus' objective has been explicitly reiterated in other international documents, notably the Addis Ababa Action

1 See for example the OECD's *Paris Declaration on Aid Effectiveness* (2005) < https://www.oecd.org/dac/effectiveness/parisdeclarationandaccraagendaforaction.htm>.

2 Emphasis added. *The Monterrey Consensus of the International Conference on Financing for Development* (2002), para 52.

DOI: 10.4324/9780429277207-7

Agenda on Financing for Development. This Action Plan emphasises 'the importance of the coherence and consistency of the international financial and monetary and trading systems in support of development',[3] and recognises 'the importance of policy coherence for sustainable development and [...] call[s] upon countries to assess the impact of their policies on sustainable development'.[4]

As if the necessity of integrating a myriad of areas such as trade, investment, finance and development was not complicated enough, additional subject-areas – importantly climate change – have recently been added to this mix. A 2017 Joint Statement by a group of development banks called on the necessity of integrating the objectives of the Paris Agreement into their own operations.[5] A 2019 OECD report supports this idea and states that,

> [i]nvesting in activities that mitigate and adapt to climate change in those environments is the only way bilateral and multilateral providers of development co-operation can carry out their mandates to support just, inclusive development and protect the natural systems that underpin life on Earth.[6]

This OECD report further recommends that development policies will be regarded as a 'catalyst' for the transition into a low carbon-based economy,[7] identifying as 'levers' for alignment tools such as capacity building, and, importantly for this chapter, also policy support (i.e. 'identifying and formulating policy and regulatory measures that provide and enabling environment'[8]) and finance.[9] This report also identifies similar problems that are highlighted throughout this study, notably that '[p]lanning for climate action is often siloed from broader development and sector plans'.[10]

This background implies that actors in the field of international development law are aware of the need to improve PCD, and address the challenge of international development through a very wide lens – notably through

3 *Addis Ababa Action Agenda of the Third International Conference on Financing for Development* (UN 2015), para 103.
4 Ibid.
5 *The MDBs' Alignment Approach to the Objectives of the Paris Agreement: Working Together to Catalyse Low-emissions and Climate-resilient Development* (n 100).
6 OECD, *Aligning Development Co-operation and Climate Action: The Only Way Forward* (OECD 2019) 13.
7 OECD, *Aligning Development Co-operation and Climate Action: The Only Way Forward* (OECD 2019) 21.
8 OECD, *Aligning Development Co-operation and Climate Action: The only Way Forward* (OECD 2019) 22.
9 OECD, *Aligning Development Co-operation and Climate Action: The Only Way Forward* (OECD 2019) 22.
10 OECD, *Aligning Development Co-operation and Climate Action: The Only Way Forward* (OECD 2019) 22.

the use of international law Home Country Measures (HCMs), such as those discussed throughout this study. Also, as reviewed below, in the more specific context of RE-related investment, policy-makers are designing development policies through the lens of PCD. As also reviewed below, these efforts are far from satisfactory and important international law HCMs are currently [being] overlooked.

A International development law? A clarification regarding this chapter's focus and limits

Before commencing with a review of relevant international law HCMs, a short clarification is required regarding this chapter's (limited) focus and coverage. This chapter addresses international development law. As is the case with other chapters in this study, the link between this field and RE-related investment is enormous in scope: the field of international development law is extremely wide and includes many relevant topics, ranging from human rights, finance, trade, investment, humanitarian aid, sustainability, education, security and more. Furthermore, the term 'international development law' is often fluid and undetermined in terms of focus and nature[11]; a characteristic that makes a concrete discussion on this area even more challenging. In light of the above, and considering this study's own limitations (notably time constraints, and the limited space of one chapter), this chapter will present a far more targeted and limited discussion.

This chapter will therefore focus on three specific areas: Official Development Assistance (ODA), the operation of development institutions, and Development Agreements. These three areas and their relevance to investment in the renewable energy (RE) sector will be discussed, and the levels of policy coherence in existing states' practice will be evaluated.

II Official Development Assistance

One of the most straightforward, and important, tools in the arsenal of international law for the attainment of the Sustainable Development Goals (SDGs) is Official Development Assistance (ODA). The use of ODA is classified as falling under the United Nations Conference on Trade and Development's (UNCTAD) fourth type of HCMs ('financial and fiscal incentives'), and is therefore of direct importance to this study. ODA funds are considered as a major channel for the transfer of resources from the Global North to the Global South[12]: in 2019 alone, OECD countries transferred 152.8

11 Rumu Sarkar, *International Development Law: Rule of Law, Human Rights & Global Finance* (OUP 2009) xvi.

12 The reader should note that ODA is not the only source of finance transferring from the developed to the developing world. The focus on ODA in this chapter is primarily due to the link with international development law, and, on the face of it, the link that ODA

billion USD,[13] including 34.62 billion USD from the US,[14] and EURO 75.2 billion from the European Union.[15]

The transfer of ODA funds takes place through a variety of international arrangements. Some of these channels are multilateral in nature and are linked to fields already discussed in this study, notably international environmental law.[16] Other arrangements are bilateral in nature: here, the role of development agreements, or even trade and investment agreements, is significant.

As discussed below, ODA funds could be used for the promotion of investment in the RE sector in many different ways. Importantly, these funds could address a variety of investment barriers, notably the high cost of capital that is inherent in RE-related investments. ODA funds could be further relied on for the purpose of capacity building, infrastructure improvement/creation, the subsidisation of financial products/guarantees (e.g. Political Risks Insurances) or, even as a direct support, financing developing countries' RE support schemes. Perhaps more controversially, ODA funds could be used as export subsidies for environmental goods and services.[17] The use of funds for the reduction of costs could indeed be done in a great many ways, all of which would increase the attractiveness of RE-related investments.

This study's objective is to examine the international laws that govern the operation of international law HCMs through the perspective of the term Policy Coherent for Development (PCD). It asks whether these international rules are designed in isolation, and what more could be done in order to create synergies and linkages between different HCMs. The following part therefore provides some background on the use (and the potential use) of ODA in the context of promoting investment in the RE sector. It will then address the international laws that govern the operation of ODA funds, and will ask whether these are designed in line with the notion of PCD. In this respect, it will look for synergies with other HCMs from other fields, but

---

funds are intended to make between climate and development goals, making it a potentially integrationist tool.

13 OECD 'Official Development Assistance (ODA)' <https://www.oecd.org/dac/financing-sustainable-development/development-finance-standards/official-development-assistance.htm>.

14 OECD 'Compare your country: Official Development Assistance 2019 – Preliminary data' <https://public.tableau.com/views/ODA-GNI_15868746590080/ODA2019?:display_count=y&publish=yes&:origin=viz_share_link?&:showVizHome=no#1>.

15 European Commission 'The European Union remains World's leading donor of Official Development Assistance with € billion in 2019' <https://ec.europa.eu/commission/presscorner/detail/en/IP_20_674>.

16 See more on this topic in Chapter 7.

17 For example, where states are sending goods/services instead of direct funds (as done with military support), although such use may not, admittedly, qualify as 'ODA' per the OECD's definition.

also with safeguards, such as the protection of human rights, the environment and more.

A ODA as a tool for the promotion of RE-related Foreign Direct Investment

There is little doubt that ODA could be used (and is being used) as an HCM for the purpose of increasing RE-related Foreign Direct Investment (FDI) in the Global South. As reviewed in Chapter 1 of this study, certain market failures are increasing the cost of investment in the RE sector. In some cases, these barriers are raising RE investments' cost considerably above the cost of equivalent investments in fossil fuel-based energy,[18] especially in the developing world.[19]

These differences are attributable *inter alia* to the cost of finance, which tends to be higher in this sector due to a number of factors, including perceived risks in investment in the RE sector (especially in the developing world), and the significant upfront investment that is required in this sector. A United Nations Development Programme (UNDP) report compared the differences between investment in RE and gas-based projects, and concluded that these can be as high as 40%.[20] Other barriers, including the subsidisation of fossil fuels,[21] are harming the competitive case for RE investment even further.

As stated in Chapter 1 of this study, the increased investment level required to achieve the commitments undertaken via the Paris Agreement, as well as this agreement's 1.5°C target, are monumental.[22] While it is likely that technological improvements will, eventually, reduce the price of RE and overcome some of these investment barriers, this day has not yet arrived, and the urgency embedded in the climate crisis implies that it may not

18 Tobias Schmidt 'Low-carbon investment risks and de-risking' (2014) 4 *Nature Climate Change* 237; Jing Hu et al. 'Barriers to investment in utility-scale variable renewable electricity (VRE) generation projects' (2018) 121 *Renewable Energy* 730, 735–736.

19 Schmidt (n 108); Gregor Schwerhoff and Mouhamadou Sy 'Financing renewable energy in Africa – key challenge of the sustainable development goals' (2017) 75 *Renewable and Sustainable Energy Reviews* 393, 394. UNDP, *De-risking Renewable Energy Investment* (UNDP 2013), 32; IPCC, *Climate Change 2014: Mitigation of Climate Change. Contribution of Working Group III to the Fifth Assessment Report of the Intergovernmental Panel on Climate Change* (CUP 2014) 1211; Janosch Ondraczek et al. 'WACC the dog: The effect of financing costs on the levelized cost of solar PV power' (2015) 75 *Renewable Energy* 888.

20 Oliver Waissbein et al. *De-risking Renewable Energy Investment. A Framework to Support Policymakers in Selecting Public Instruments to Promote Renewable Energy Investment in Developing Countries* (2013 UNDP) 34.

21 Doug Koplow 'Defining and measuring fossil fuel subsidies' in Jakob Skovgaard and Harro van Asselt (eds.) *The Politics of Fossil Fuel Subsidies and Their Reform* (CUP 2018), 32.

22 IRENA/IEA *Perspectives for the Energy Transition: Investment Needs for a Low-Carbon Energy System* (IEA/IRENA 2017).

arrive fast enough for the planet, or for the millions who are currently living without access to energy.

These issues are not new and policy-makers are aware of them. Indeed, most countries have adopted RE support schemes for the purpose of mitigating these barriers and encouraging increased investment in the RE sector.[23] However, support schemes imply 'support', which in turn implies the allocation of resources. Such resources are not always available in developing states, where more urgent concerns compete for a much smaller pool of funds.[24] Matters of environmental justice and global responsibility reduce even further developing countries' political will to 'subsidise' investment in the RE sector.[25]

One widely supported answer to this financial gap is derived from the principle of common but differentiated responsibilities.[26] In short, it is suggested that the Global North will bear some of the financial burden that is inherent in the developing world's transition into a green economy. Today, this view represents the mainstream thinking in global climate negotiations, even if its concrete implementation at times remains disputed.

The financial support provided by the industrialised world is arriving in different forms, including through ODA, defined by the OECD as 'government aid designed to promote the economic development and welfare of developing countries'.[27] According to the OECD's definition, a significant part of the funds transferred for the purpose of supporting RE-related investment is regarded as ODA.[28]

As stated by de Sépibus, the use of ODA funds is progressively changing[29]:

> The delivery of ODA has undergone important changes over the past two decades. Notably, the past decade has witnessed a massive increase

23 According to the OECD, by 2015, 145 states have adopted national support schemes in order to promote the RE. See: OECD, *Overcoming Barriers to International Investment in Clean Energy* (OECD 2015) <http://www.oecd-ilibrary.org/environment/green-finance-and-investment_24090344> 19.

24 See data on number of states' support schemes, according to their levels of income in REN21, *Renewables 2015: Global Status Report* (REN21 2015) Table 3, 99.

25 See the meaning, history and origins of the principle of common but differentiated responsibilities under the climate change regime in Jutta Brunnée and Charlotte Streck 'The UNFCCC as a negotiations forum: Towards common but more differentiated responsibilities' (2013) 13(5) *Climate Policy* 589.

26 The principle of common but differentiated responsibilities can be found in almost all of the UNFCCC's major legal documents, *inter alia* the Kyoto Protocol, the Copenhagen financial pledges, the Paris Agreement and, as reviewed below, the different climate funds.

27 OECD 'OECD Net ODA' <https://data.oecd.org/oda/net-oda.htm>.

28 See OECD 'Is It ODA?' (November 2008) *Factsheet* <https://www.oecd.org/dac/stats/34086975.pdf>.

29 Joëlle de Sépibus 'Green climate fund: How attractive is it to donor countries?' (2015) 9(4) *Carbon & Climate Law Review* 298, 309.

> of multi-bi-financing and more recently the focus has been placed on value for money, the leverage of private funds and the involvement of local as well as civil society actors.

This change is certainly in line with the notion of PCD: it implies the utilisation of linkages and synergies with other policies, such that will increase ODA investments' value for money. This realisation is especially important in the case of RE investment. The OECD's Better Policies for Sustainable Development document supports the utilisation of ODA as a tool for leveraging further investment (rather than directly funding investment), stressing that 'official development assistance (ODA) remains essential in creating an enabling global environment for green growth while supporting specific measures in developing countries'.[30] In this context this organisation specifically mentions support that is aimed at facilitating RE-related FDI, notably capacity building and the facilitation of international trade in environmental goods and services.[31]

A different OECD report further provides recommendations on the effective use of ODA, emphasising its role in a much wider, holistic framework of investment facilitation policies[32]:

> It is clear that if the opportunities offered by the transition to a green economy are to be realised, all nations need to be brought on board – including the refocusing of official development assistance (ODA) to support green economy and growth in developing countries. For the poorer countries, however, big obstacles remain; these range from complex issues – such as improved governance – to simple realities, such as the currency exchange risks that dampen the enthusiasm of investors, no matter how vast the opportunity offered by a wind or solar resource project may be. *If carefully and strategically deployed, ODA could play a catalytic role in overcoming some of these barriers, as could implementing and financially supporting a green climate fund.*

The above implies that the use of ODA funds should not be designed in isolation. Rather, ODA should be understood as part of a much wider mechanism; one that includes different rules on a myriad of issues, such as trade, investment and insurance, all designed holistically in a mutually supportive manner in order to achieve the grand objective of increasing sustainable RE investment in the Global South.

30 OECD, *Better Policies for Sustainable Development* (OECD 2016) 205.

31 OECD, *Better Policies for Sustainable Development* (n 30) 205.

32 OECD, *Development Co-operation Report 2012: Lessons in Linking Sustainability and Development* (OECD Publishing 2012). http://dx.doi.org/10.1787/dcr-2012-en at169. Emphasis added.

B ODA and climate change: critical views

The allocation of ODA funds for climate-related purposes has been met with some criticism, as well as with scepticism. According to one oft-heard criticism, funds that are allocated to climate change mitigation are often diverted from existing foreign aid budgets, and are not (as required by the Cancun Decision[33]) 'new and additional'.[34] Some point to a rather cynical definition of "climate"-related ODA funds: Homi Kharas provides examples of ODA funds directed at projects such as metro systems, dam projects and agricultural projects, which were re-branded as 'climate-related support' due to their somewhat indirect impact on climate change.[35]

Kaul questions whether the transfer of climate-related ODA funds should be regarded as a 'transfer' at all. Climate funds, she explains, are designed to finance public goods, enjoyed not only by the recipient state, but rather by the international community as a whole. As such, she argues, '[s]uch finance constitutes payment for a service, and is not a transfer'.[36] A somewhat related argument is made by Ryan, who questions the compatibility of climate mitigation projects with the definition of ODA:

> If a project qualifies as "principally" about climate mitigation, which it needs to be under the OECD Rio markers to count as 100% climate finance, then it is hard to argue that the main objective is the economic development and welfare of that particular developing country.[37]

Others are also adding to this growing pool of criticism the fact that most climate-related funds flow towards middle-income countries, leaving low-income countries vulnerable.[38] While developing countries view these transfers as, at least partially, a compensation for past emissions by developed countries, developed countries are looking for a cost-effective use of their

33 UNFCCC, Decision 1/CP.16 (2010) un Doc FCCC/CP/2010/7/Add.1, para 95.

34 Fiona Ryan 'Separating climate finance and ODA' (24 April 2019) *reliefweb* <https://reliefweb.int/report/world/separating-climate-finance-and-oda>; Homi Kharas 'Aid and climate finance' (2015) *Brookings* <https://www.brookings.edu/wp-content/uploads/2016/07/aid-climate-finance-kharas-1.pdf>.

35 Kharas (n 34).

36 Inge Kaul, Robin Davies, Robert Glasser, Michael Gerber and Luca Etter 'Policy debate: Financing the SDGs: Global vs local public goods' (2015) International Development Policy, Revue internationale de politique de développement <https://journals.openedition.org/poldev/2068#tocto3n1>.

37 Ryan (n 34).

38 Kharas (n 34) 26; Giorgio Gualberti and Lorcan Lyons, *Missing the Mark: Gaps and Lags in Disbursement of Development Finance for Energy Access* (Sustainable energy for all 2017) <https://www.afdb.org/fileadmin/uploads/afdb/Documents/Publications/Missing_the_Mark_Gaps_and_Lags_in_Disbursement_of_Development_Finance_for_Energy_Access.pdf> 8.

funds. They (developed nations) therefore prefer to target middle-income countries, where investment risks and conditions are favourable.[39]

These difficult issues are important and deserve further attention. At the same time, it must be acknowledged that many of these issues are in fact part and parcel of the hard trade-offs implied by the act of policy integration.

For example, the criticism over the (lack of) 'additionally' is without a doubt concerning; where ODA funds are indeed simply re-directed towards climate-related projects (rather than added as entirely new funds), it means that less funds are made available for other important causes. Nevertheless, as discussed in Chapter 2 of this study, the political decision to prioritise the climate over other public goals is hardly unique to the context of ODA and renewables; this trade-off is present in almost *any aspect* of climate policy. This prioritisation of resources undeniably represents a political decision, and if the goal of addressing climate change is to be made a priority, there is a price that must be paid. There is never enough money to go around, and the only way out of these budgetary trade-offs will be to increase developed nations' financial contributions – indeed, a political commitment that developed nations have made in the past, but, so it seems, are struggling to deliver.

Furthermore, it seems unfair to claim that these budgetary choices are *always* between climate-related objectives and other policy goals. The two are not necessarily mutually exclusive, and one must not forget the range of non-climate related (and very much development-oriented) benefits that are inherent in RE projects in the developing world.[40] At least in some cases, the diversion of funds towards RE projects does not mean less attention for objectives such as education, health or gender equality. A hospital, after all, can hardly function without electricity.

The second criticism, concerning the destination of climate-related ODA (often to middle-income states), reflects no less than a full-frontal clash between SDG 7 and SDG 13 (climate action): investment based on better value for money (in terms of emission reductions) vs. investment based on need. From a purely climate perspective, a tonne of CO_2 reduced in a middle-income country is as effective as a tonne of CO_2 reduced in a least developed country. The prioritisation of middle-income countries is understandable from this narrow perspective: a similar financial contribution will simply

39 Kharas (n 34) 27.

40 See review in Chapter 1 of this study, on the positive impact that RE project could have on a range of SDGs. See also Gregor Schwerhoff and Mouhamadou Sy 'Financing renewable energy in Africa – key challenge of the sustainable development goals' (2017) 75 *Renewable and Sustainable Energy Reviews* 393; Nadia Ouedraogo 'Opportunities, barriers and issues with renewable energy development in Africa: A comprehensible review' (2019) 5 *Current Sustainable/Renewable Energy Reports* 1; Nicola Cantore et al. 'Promoting renewable energy and energy efficiency in Africa: A framework to evaluate employment generation and cost effectiveness' (2017) 12 *Environmental Research Letters* 035008.

achieve a better result. Here political decisions will also have to be made in order to balance and prioritise conflicting interests. This study's objective is indeed to assist with this balancing activity, by offering relevant legal instruments and principles that will improve the case for investment in the developing world.

C Climate funds and PCD

The Inter-governmental Panel on Climate Change (IPCC) presented (now somewhat dated) data on climate-related ODA: together, the seven multilateral development banks (MDBs) distributed 24.1 billion USD in 2011, and 26.8 billion USD in 2012.[41] A more recent assessment by the OECD estimates bilateral climate-related development finance in 2016 at 30 billion USD.[42] The transfer of ODA to support developing countries' transition into a low-carbon economy is also expected to increase in the future as the 2015 United Nations Framework Convention on Climate Change (UNFCCC) Paris Agreement instructs the developed world to transfer financial assistance (such as ODA) in order to support developing states' attempts to meet the objectives of the UNFCCC.[43] It is not clear however whether such increase has been noticeable in practice,[44] or whether it will even be genuine, additional and new (see criticism above).

A significant part of ODA transfers are administered through public international funds. The landscape of this area is wide, diverse and almost inherently fragmented: an IPCC report identified approximately 50 of these international funds,[45] as well as numerous other public finance sources (including 45 carbon market funds) and thousands of private sources (around 6,000 private funds).[46] For reasons of time, scope and space, the author will not review this myriad of funds. A brief review of the main public funds –

41 IPCC Sujata Gupta et al. (eds) 'Chapter 16: Cross-cutting investment and finance issues' in *Climate Change 2014: Mitigation of Climate Change* (IPCC 2014) 1215.

42 OECD, *Climate-related Development Finance in 2016* (OECD 2017) <http://www.oecd.org/dac/stats/climate-change.htm>.

43 See The Paris Agreement (adopted 12 December 2015, entered into force 4 November 2016) (Paris Agreement) Art 9.

44 A 2019 OECD report states that

> Between 2014 and 2017, the share of climate-related Official Development Assistance reported to the OECD DAC remained stable at around 20–21%, after a slight increase between 2013 and 2014. During this same period, the share of multilateral climate finance in total multilateral outflows to ODA-eligible countries grew from 18% in 2013 to 28% of total multilateral outflows in 2017.

OECD, *Climate finance Provided and Mobilised by Developed Countries in 2013-17* (OECD 2019) 10.

45 IPCC 'Chapter 16: Cross-cutting investment and finance issues' in *Climate Change 2014: Mitigation of Climate Change* (IPCC 2014) 1228.

46 Ibid.

the Green Climate Fund (GCF) and the Global Environmental Facility (GEF) – is nevertheless important.

i The Green Climate Fund

The Green Climate Fund (GCF) was established in 2010 and became operative in 2015.[47] The GCF is operated under instructions from the UNFCCC's Conference of Parties (COP), and is governed by a Board with equal representation from developed and developing countries.[48] The GCF offers different financial instruments, including grants, loans, equity investments and guarantees,[49] and is financed almost exclusively by developed, investment exporting countries.[50] The resources of the GCF are distributed via MDBs, as well as through a list of public and private accredited entities.[51]

The GCF's relevance to this study is noticeable on several levels. To begin with, the GCF's objectives are similar to that of this study – i.e. to address investment barriers in order to 'tip' investors' decisions towards investment in the RE sector. In this respect, the GCF's Governing Instrument states that '[f]inancing will be tailored to cover the identifiable additional costs of the investment necessary to make the project viable'.[52] Furthermore, the GCF is explicitly aligned towards the promotion of RE investment: one of the fund's four declared 'mitigation strategic impacts' is 'energy generation and access',[53] effectively mirroring the objective of SDG 7. Another related, relevant feature is the GCF's Private Sector Facility. The Private Sector Facility was established for the purpose of scaling up private sector

47 For more details about the fund and its operation see: Liane Schalatek et al. 'The green climate fund' (2015) *Climate Finance Fundamentals* <https://www.odi.org/sites/odi.org.uk/files/odi-assets/publications-opinion-files/10066.pdf>.

48 Green Climate Fund, *Governing Instrument for the Green Climate Fund* (UNFCCC 2011) <https://www.greenclimate.fund/sites/default/files/document/governing-instrument.pdf> para 9.

49 See a review of projects supported by the fund so far in Green Climate Fund, *Project Briefs* (Green Climate Fund 2015) <http://www.greenclimate.fund/documents/20182/194568/GCF_Project_Briefs_2015.pdf/b3cb6cd3-cac4-409f-92e7-028ad2fb902b>.

50 Interestingly, several middle-income countries have pledged to contribute funds to the GCF, although these funds are relatively small and are understandably dwarfed by developed countries' contributions. See Green Climate Fund 'Status of Pledges and Contributions made to the Green Climate Fund' (31 July 2020) <https://www.greenclimate.fund/sites/default/files/document/status-pledges-irm-gcf1_0.pdf>.

51 List of accredited entities can be found here: <http://www.greenclimate.fund/documents/20182/114261/20160516_-_GCF_List_of_Accredited_Entities.pdf/e09bb9b3-9730-4adc-bca9-ff32739ecae8>.

52 Green Climate Fund, *Governing Instrument for the Green Climate Fund* (n 48) para 54.

53 Green Climate Fund, *Investment Opportunities for the Green Climate Fund: GFC's Role and Impact within the Climate Finance Ecosystem* (GCF 2015) 2.

investments, making it a classic HCM (especially as its funds arrive from investment exporting countries).[54]

The GCF's link with development objectives is clear: the GCF's secretariat describes this fund as the 'world's largest dedicated fund helping developing countries reduce their greenhouse gas emissions and enhance their ability to respond to climate change'.[55] The GCF's Governing Instrument repeats this line, stating that (under 'objectives and guiding principles')[56]:

> The Fund will contribute to the achievement of the ultimate objective of the United Nations Framework Convention on Climate Change (UNFCCC). In the context of sustainable development, the Fund will promote the paradigm shift towards low emission and climate-resilient development pathways by providing support to developing countries to limit or reduce their greenhouse gas emissions and to adapt to the impacts of climate change, taking into account the needs of those developing countries particularly vulnerable to the adverse effects of climate change.

The GCF, it seems, is a significant tool for the promotion of RE-related investments in the developing world, and as such it deserves closer scrutiny.

THE GREEN CLIMATE FUND AND POLICY COHERENCE FOR DEVELOPMENT

With respect to PCD, the GCF's Governing Instrument includes a commitment according to which the GCF shall operate 'in the context of sustainable development'.[57] In this respect, Paragraph 3 of the GCF's Governing Instrument adds that '[t]he Fund will strive to maximize the impact of its funding for adaptation and mitigation, and seek a balance between the two, while promoting environmental, social, economic and development co-benefits and taking a gender-sensitive approach'.[58]

These instructions imply a commitment to a PCD-oriented approach and an attempt to link this fund's operation to a variety of policy objectives. Indeed, the GCF's activity in recent years demonstrates some level of commitment to this goal: notably, a line of guidelines and policies

54 Views concerning the success and effectiveness of the Private Sector Facility in attracting private investment differ. See Rajesh Eralil, Mariama Williams and Dianyi Li 'Flirting with the private sector: The GCF private sector facility—achievements, challenges and constraints in engaging the private sector' (2020) 23 *South Centre: Climate Policy Brief* <https://www.southcentre.int/wp-content/uploads/2020/03/Climate-PB-23.pdf>.

55 GCF <https://www.greenclimate.fund/about>.

56 Green Climate Fund, *Governing Instrument for the Green Climate Fund* (n 48) para 2.

57 Green Climate Fund, *Governing Instrument for the Green Climate Fund* (n 48) para 2.

58 Green Climate Fund, *Governing Instrument for the Green Climate Fund* (n 48) para 2.

have been adopted, on issues such as gender equality,[59] indigenous people,[60] stakeholder engagement[61] and, importantly, also environment and social matters.[62] For example, the GCF Environmental and Social Policy states[63]:

> Human rights. All activities supported by GCF will be designed and implemented in a manner that will promote, protect and fulfil universal respect for, and observance of, human rights for all recognized by the United Nations. GCF will require the application of robust environmental and social due diligence so that the supported activities do not cause, promote, contribute to, perpetuate, or exacerbate adverse human rights impacts; and
>
> Biodiversity. All GCF-financed activities will be designed and implemented in a manner that will protect and conserve biodiversity and critical habitats, ensure environmental flows of water, maintain the benefits of ecosystem services, and promote the sustainable use and management of living natural resources.

Further, the GCF has created certain safeguards to ensure the protection of all these interests, including a grievance mechanism (discussed elsewhere in this study[64])[65] and a requirement to rely on public participation.[66]

In theory, the GCF's efforts seem comprehensive and impressive. These developments, however, are relatively recent and may require certain fine-tuning: indeed, an independent evaluation of the GCF's Environmental and Social Safeguards (ESS) and the Environmental and Social Management System identified certain shortcomings with respect to the GCF's implementation of all of these mechanisms.[67] These

59 Green Climate Fund, *Gender Policy* (GCF 2018) <https://www.greenclimate.fund/sites/default/files/document/gcf-gender-policy.pdf>.

60 Green Climate Fund, *Indigenous Peoples Policy* (GCF 2018) <https://www.greenclimate.fund/sites/default/files/document/ip-policy.pdf>.

61 GCF, Decision GCF/B.08/45, Annex XIV: Initial best-practice options for country coordination and multi-stakeholder engagement <https://www.greenclimate.fund/sites/default/files/decision/b08/decision-b08-10-b08-a14.pdf>.

62 Green Climate Fund, *Environment and Social Policy* (GCF 2018) <https://www.greenclimate.fund/sites/default/files/document/environment-social-policy.pdf>. The GCF's Board adopted further a line of procedures on environmental and social safeguards, see for example.

63 Green Climate Fund, *Environment and Social Policy* (n 62) Part IV.

64 See Chapter 7, in this study.

65 Green Climate Fund, *Procedures and Guidelines of the Independent Redress Mechanism* (GCF 2019) <https://www.greenclimate.fund/sites/default/files/document/procedures-guidelines-irm.pdf>; see also Green Climate Fund, *Environment and Social Policy* (n 62) part 7.3.

66 Green Climate Fund, *Environment and Social Policy* (n 62) Part 7.2.

67 Green Climate Fund, *Independent Evaluation of the Green Climate Fund's Environmental and Social Safeguards (ESS) and the Environmental and Social Management*

included 'important gaps regarding human rights, gender, and equity concerns'. *Inter alia*[68]:

> The GCF has not adopted any guidance on how to screen and assess potential adverse effects on human rights, either for itself or for AEs [Accredited Entities, AK]; Its Gender Policy does not adhere to international standards, and there is no guidance on what constitutes 'consent' among others. [...]
>
> The current environmental and social management system does not focus adequately on generating positive, measurable social and environmental improvements as co-benefits that are otherwise a *sine qua non* for climate investments.

And,

> The GCF does not require or provide guidelines for how funded projects/investments should report on social and environmental outcomes and performance as well as co-benefits and, compared with other climate funds, has less stringent reporting requirements on safeguard implementation.

Also, with respect to reporting and monitoring, the independent review found that '[t]he GCF does not adequately monitor ESS compliance, social and environmental outcomes, and/or co benefits of funded projects/programmes'.[69] A reliance on self-reporting (a criticism that was made also with respect to the Kyoto Protocol[70]) was also mentioned as a weakness.[71]

As for coherence/synergies with other international law HCMs, Paragraph 33 of the Governing Instrument prescribes that '[t]he Fund shall operate in the context of appropriate arrangements between itself and other existing funds under the Convention, and between itself and other funds, entities, and channels of climate change financing outside the Fund'.[72] On the face of it, this instruction mostly seems to focus on the narrower perception of coherence – between the GCF and other financing entities alone, leaving wider interactions with entities that are indirectly related to finance – such

System (ESMS) (2020) GCF/B.25/07 <https://www.greenclimate.fund/sites/default/files/document/gcf-b25-07.pdf>.

68 Green Climate Fund, *Independent Evaluation of the Green Climate Fund's Environmental and Social Safeguards (ESS) and the Environmental and Social Management System* (n 68) xx–xxi.

69 Green Climate Fund, *Independent Evaluation of the Green Climate Fund's Environmental and Social Safeguards (ESS) and the Environmental and Social Management System* (n 68) xxii.

70 See review in the Chapter 7, in this study.

71 Green Climate Fund, *Independent Evaluation of the Green Climate Fund's Environmental and Social Safeguards (ESS) and the Environmental and Social Management System* (n 68) xxiii.

72 Green Climate Fund, *Governing Instrument for the Green Climate Fund* (n 48) para 33.

as the World Trade Organization (WTO), for example – outside of the scope of this instruction. On the other hand, Paragraph 33 also includes the term 'entities', explicitly distinguished from 'other funds' or 'channels of climate change financing'. This reference could be understood as implying a wider mandate, in order to cooperate with other institutions such as the Multilateral Investment Guarantee Agency (MIGA) or the WTO. Indeed, Paragraph 34 adds that:

> The [GCF] Board will develop methods to enhance complementarity between the activities of the Fund and the activities of other relevant bilateral, regional and global funding mechanisms and institutions, to better mobilize the full range of financial and technical capacities.

At the time of writing, and to the best of the author's knowledge no efforts to link the operation of the GCF with trade, insurance or investment HCMs are taking place. It seems that for the time being, the GCF's focus is on improving its linkages with other financing institutions, and, importantly, environmental and human rights institutions – a monumental task on its own. Wider interactions (e.g. with trade and investment instruments) – while arguably falling within the GCF mandate – are not yet on the table.

The GCF would do well to explore these wider interactions, especially as its Governing Instrument explicitly instructs it to 'strive to maximize the impact of its funding for adaptation and mitigation'.[73] These tools, as explained in other chapters in this study, will improve the GCF's investments 'value for money' by addressing a variety of investment barriers. A loan or a grant, on its own, is useful, but will be regarded as far more attractive when delivered with a suite of protection rules and guarantees that reduce the project's risks, or with trade facilitation rules that increase the ease of doing business. The GCF will do well in this respect to think beyond the mere transfer of funds.

ii Global Environmental Facility

The GEF was established in 1991 in order to support the finance of a variety of international environmental agreements. The GEF was designated in the UNFCCC itself (1992)[74] as its original financing mechanism.[75] According to the GEF's reporting, since its inception a total of 5.2 billion USD has been invested in 167 countries, for 839 projects related to climate adaptation and mitigation.[76]

73 Green Climate Fund, *Governing Instrument for the Green Climate Fund* (n 48) para 2.
74 See Article 21(3) of the UNFCCC.
75 See Article 11(1) of the UNFCCC.
76 GEF, *Report of the Global Environment Facility to the Twenty-first Session of the Conference of the Parties to the United Nations Framework Convention on Climate Change* (GEF 2015) 2.

Much of this finance has been invested in order to foster and scale up the activity of the private sector via the deployment of a variety of financial instruments.[77] Despite these efforts, GEF's engagement with the private sector to date has not been satisfactory.[78] The need to improve this aspect was openly admitted in the GEF's most recent Programming Directions (GEF-7), and the possibility of forming an advisory group of experts to address this element is being discussed.[79]

The link between the GEF's activity and development, as well as with investment in the RE sector, is straightforward. GEF funds are contributed by a list of donor countries, most of which are developed nations,[80] making it a clear HCM. Indeed, contributions to the GEF are considered as ODA.[81] In this respect, the GEF's Programming Directions state[82]:

> Building on the GEF-6 Focal Area Strategy and in alignment with UNFCCC COP guidance, the GEF-7 Climate Change Focal Area Strategy aims to support developing countries to make transformational shifts towards low emission and climate-resilient development pathways.

Like the GCF, and in line with the manner in which ODA is increasingly perceived by states (mostly developed states), the GEF increasingly stresses the importance of value for money for its investment, underscoring the need to 'ensure the cost-effectiveness of its [the GEF's, AK] activities'.[83]

77 See for example the GEF non-grant pilot <https://www.thegef.org/gef/NGI>.

78 GEF, *GEF-7 Programming Directions* (2018) GEF/R.7/19 April 2, <https://www.thegef.org/sites/default/files/publications/GEF-7%20Programming%20Directions%20-%20GEF_R.7_19.pdf> 131-136.

79 It is stated in the GEF-7 that this expert group

> will be called on to help identify strategic investment trends and opportunities for platforms for aggregation and scaling, best use for specific types of non-grant instruments, and potential investment categories/platforms (e.g., sustainable fisheries, agro/forestry, chemicals and waste) where innovation is needed.

GEF-7 Programming Directions (n 78) 136.

80 For a list of donor countries, see on GEF's website <https://www.thegef.org/partners/participants>.

81 Please see Note prepared by the OECD secretariat which lists GEF as one of the institutions 'core contributions to which may be reported as official development assistance, either in whole or in part'. OECD Secretariat 'Is It ODA?' (May 2007) *Factsheet* <https://www.oecd.org/dac/stats/34086975.pdf>.

82 GEF, *GEF-7 Programming Directions* (n 78) 36.

83 See in Art I(5) of the GEF's Instrument for the Establishment of the Restructured Global Environmental Facility'. The 'Instrument' is essentially a treaty between the relevant states, which regulates the GEF's activity, available online: GEF 'Instrument for the Establishment of the Restructured GEF' (2 March 2015) <https://www.thegef.org/documents/instrument-establishment-restructured-gef>.

GLOBAL ENVIRONMENTAL FACILITY AND POLICY COHERENCE FOR DEVELOPMENT

There is no doubt that efforts to improve PCD are taking place within the GEF. The GEF-7 Programming Directions explicitly address the GEF's role as a facilitator of investment, notably through increased policy coherence[84]:

> The GEF-7 Climate Change Focal Area Strategy is specifically designed to be complementary to programming by the GCF and other climate funds, based on the GEF's unique role in the global environmental finance architecture to lay the foundation for enhanced climate action, namely by 1) *harnessing synergies across the different focal areas in line with an integrated approach to generate multiple global environmental benefits*; and 2) building on the GEFs long-standing track record of driving innovation and funding demonstration and pilot activities that are too early in the market adoption chain to be within the reach of other providers of environmental finance.

And[85]:

> [GEF's strategy] does not prioritize direct support for large-scale deployment and diffusion of mitigation options with GEF financing only. Rather, GEF-7 resources should be utilized to reduce risks and enhance enabling environments, so that the results can facilitate additional investments and support by other international financing institutions, the private sector, and/or domestic sources to replicate and scale up in a timely manner.

Further, a GEF Background Note states on this point[86]:

> To take advantage of the GEF's comparative advantage in a complex financial architecture, resources will seek to reduce risks and enhance enabling environments to facilitate the replication and scale up of additional investments by other international financing institutions, the private sector, and domestic sources.

On the face of it, these statements reflect a very strong reliance on the concept of PCD: the GEF sees itself as a completing, leveraging tool; one

84 GEF, *GEF-7 Programming Directions* (n 78) 36.
85 GEF, *GEF-7 Programming Directions* (n 78) 38.
86 GEF Secretariat 'GEF Assembly Background Note: Innovation in Clean Energy' (2018) <https://www.thegef.org/publications/gef-assembly-background-note-innovation-clean-energy>.

that will be used to enhance the wider investment environment and complete the efforts of other institutions, *inter alia*, through the promotion of synergies.[87]

Despite the above-stated PCD-oriented language, and despite explicit recognition that more should be done in order to increase investment from the private sector, the GEF's governing documents address synergies with other institutions in a rather limited manner and only in very specific areas. Markedly, the GEF is increasingly cooperating with other climate funds, notably the GCF. The GEF's report to the UNFCCC dedicates a section to 'complementarity in climate finance', in which it describes its efforts to increase links with the GCF, including efforts 'to minimize overlapping support, and to enhance coordination'.[88] While to date this cooperation mostly includes different forms of information exchange (e.g. co-hosting of events, increased dialogue and invitations to attend the other institutions' meetings), in the future it seems that these efforts will also expand to activities such as coordinated engagement.[89]

While efforts at improving coherence with other HCMs seem limited, the same cannot be said concerning the GEF's aspiration to impose necessary safeguards, i.e. to link its finance with subject-areas such as environmental protection and human rights. Notably, the GEF adopted a Policy on Environmental and Social Safeguards[90] (adopted in 2011, and revised on December 2018), which was later supplemented with accompanying guidelines.[91] The GEF's Social and Environmental Policy applies to all of the GEF's funded projects. It is aimed at

> avoiding, preventing, minimizing, mitigating, managing, offsetting or compensating any adverse impacts that GEF financed projects and programs may have on people or the environment throughout the project

87 GEF, *GEF-6 Programming Directions* (2014–2018) (GEF 2014) <https://www.thegef.org/gef/sites/thegef.org/files/webpage_attached/GEF6_programming_directions_final_0.pdf> 52 and 54 (see Box 2 in this page).

88 GEF, *Report of the Global Environment Facility to the Twenty-Fifth Session of the Conference of the Parties to the United Nations Framework Convention on Climate Change* (26 August 2019) <https://www.thegef.org/sites/default/files/documents/gef_report_unfccc_cop25.pdf> para 33.

89 See more on the GEF/GCF plans for coordinated engagement in GCF/GEF 'Coordinated Engagement by Green Climate Fund and Global Environmental Facility' (2018) <https://assembly.thegef.org/sites/default/files/documents/GCF-GEF%20partnership%20concept%20agenda%20for%2026%20June%202018%20rev.pdf>.

90 GEF, *Policy on Environmental and Social Safeguards* (2018) SD/PL/03 <https://www.thegef.org/sites/default/files/documents/gef_environmental_social_safeguards_policy.pdf>.

91 GEF, *Guidelines on GEF's Policy on Environmental and Social Safeguards* (2019) SD/GN/03 <https://www.thegef.org/sites/default/files/documents/guidelines_gef_policy_environmental_social_safeguards.pdf>.

> or program cycle; thereby enhancing the environmental and social outcomes of such projects and programs.[92]

The GEF's Social and Environmental Policy is based on nine minimum standards, including environmental and social assessments; grievance mechanisms; bio-diversity; indigenous people; cultural heritage and more. Interestingly, while the Policy does not mention the term 'human rights', it does refer to several fundamental human rights including of those communities who have been relocated due to GEF's projects[93] as well as indigenous people,[94] and others.[95]

Other relevant policies that reflect the GEF's aspiration to improve its PCD through safeguards include GEF's policy on gender equality (2019) and on indigenous people (2012).[96] These policies are based on a line of key principles such as mainstreaming; reliance on public participation and respect for human rights[97]; as well as safeguards in the form of compliance, grievance and dispute settlement mechanisms.[98]

It can be seen that GEF's policy-makers are aware of the need to increase PCD and explore linkages with other institutions and policy-areas. It is equally clear that these efforts are limited to very specific areas, and tools from fields such as trade, investment or insurance are entirely ignored. Indeed, the author of this study did not find any reference to international law HCMs from these areas.

While the improved focus on safeguards must naturally be seen as a positive development, the limited view in which HCMs are addressed within the GEF is problematic. The fact that the efforts to improve linkages with other HCMs are focussed exclusively on the climate finance universe reflects

92 GEF, *Policy on Environmental and Social Safeguards* (n 90) para 2.

93 See minimum standard 4 in GEF, *Policy on Environmental and Social Safeguards* (n 90).

94 See minimum standard 5 in GEF, *Policy on Environmental and Social Safeguards* (n 90).

95 See for example See minimum standard 8 on labour standards, in GEF, *Policy on Environmental and Social Safeguards* (n 90).

96 See GEF's policies on this fund's website, here <https://www.thegef.org/documents/policies-guidelines>.

97 GEF, *Principles and Guidelines for Engagement with Indigenous Peoples* (GEF 2012) 13–15; GEF, *Policy on Gender Equality* (GEF 2019) 6–7.

98 GEF, *Principles and Guidelines for Engagement with Indigenous Peoples* (n 97) 21. Interestingly, while referring indigenous people's complaints 'as a first step' to local courts, the GEF takes an active role in facilitating complaints:

> the gef conflict resolution commissioner and the gef indigenous Peoples focal point will be available and actively involved in complaints brought forward by indigenous Peoples to the gef that fail to be adequately addressed at the local, country, or gef Partner agency level.

GEF, *Principles and Guidelines for Engagement with Indigenous Peoples* (n 97) 21. With respect to compliance mechanism, see GEF, *Policy on Gender Equality* (n 97) 9.

both a narrow expertise and a lack of interest to expand the tools that are available.

There are many possible explanations for the gap between the recognition of the importance of PCD on the one hand, and the limited approach to PCD on the other. These may include limited available expertise, lack of (a state's) political will to engage with other regimes through the GEF or possibly a mere lack of ambition and leadership, as the GEF's role in the climate finance universe has been reduced in recent years. In any event, in light of the GEF's experience in administering funds for different regimes and a wide variety of actors, one would have expected the GEF's operation to be more in tune with the need to increase policy coherence.

III Development institutions

Another element that should be addressed in the discussion on international development law and investment in the RE sector is the activity of international development institutions, and notably MDBs. These institutions are all creatures of international law, and their influence on ODA-related HCMs is significant. While not regarded as HCMs in their own right, they can be regarded as administrators and facilitators of HCMs, whose operation is complementary, and crucial, to the efforts of states. As such, they deserve a brief discussion in this chapter.

The role that international institutions such as the World Bank and other regional MDBs play in achieving the SDGs, and more specifically in increasing the flow of climate-related investment towards the Global South, is widely acknowledged,[99] including by the development banks themselves.[100] These institutions' support of RE-related investment is threefold[101]: (1) as a source of (affordable) finance; (2) as a tool for mobilising further, mostly private finance (*inter alia* by financing de-risking activities); and (3) by providing technical assistance and supporting governments' market and

99 See for example in the Addis Ababa Action Agenda of the Third International Conference on Financing for Development (UN 2015), para 44; Amar Bhattacharya, Homi Kharas, Mark Plant and Annalisa Prizzon 'The new global agenda and the future of the multilateral development bank system' (2018) <https://think-asia.org/bitstream/handle/11540/7984/epg_paper_on_future_of_mdb_system_jan301.pdf?sequence=1>; Bjarne Steffen and Tobias Schmidt 'The role of public investment & development banks in enabling or constraining new power generation technologies' (2017) IEEE <https://ieeexplore.ieee.org/stamp/stamp.jsp?tp=&arnumber=7981949>.

100 *The MDBs' alignment approach to the objectives of the Paris Agreement: Working together to catalyse low-emissions and climate-resilient development* (2017) <http://pubdocs.worldbank.org/en/784141543806348331/Joint-Declaration-MDBs-Alignment-Approach-to-Paris-Agreement-COP24-Final.pdf>.

101 OECD, The World Bank, UN Environment (n 103) 101.

regulatory reforms.[102] The OECD's Secretariat has defined these roles as 'critical'[103]: 'These institutions', it is explained, 'leverage finance from capital markets due to their strong credit ratings and the backing of their shareholder governments, and in turn provide financing to support development outcomes'.[104] Bhattacharya et al. mention in this respect also MDBs' 'experiences and practices gained from 70 years of development cooperation',[105] allowing them to facilitate *inter alia* 'the mobilization of financing from other sources, especially from the private sector, and an ability to address impediments and de-risk programs and projects'.[106]

Development institutions are aware of their role and its importance. In recent years, MDBs have made significant commitments to scale up their investment in climate finance. Indeed, these investments are rising on a yearly basis,[107] with most MDBs meeting their climate finance targets.[108]

At the same time, it is clear that the operation of development institutions could be improved. As discussed below efforts by MDBs to mobilise private finance have been criticised for failing to meet their 'from billions to trillions' leveraging agenda,[109] attracting less than $1 in private investment for every $1 invested.[110] As mentioned by Thwaites, '[i]t's time for MDBs to re-evaluate the "Billions to Trillions" agenda and explore whether other approaches may be more effective at promoting private investment in climate action'.[111]

102 OECD, The World Bank, UN Environment (n 103) 101; Amar Bhattacharya, Homi Kharas, Mark Plant and Annalisa Prizzon (n 99) 7.

103 OECD, The World Bank, UN Environment, *Financing Climate Futures: Rethinking Infrastructure* (OECD Publishing 2018) <https://www.oecd-ilibrary.org/environment/financing-climate-futures_9789264308114-en> 100.

104 OECD, The World Bank, UN Environment (n 103) 100.

105 Amar Bhattacharya, Homi Kharas, Mark Plant and Annalisa Prizzon (n 99) 7.

106 Amar Bhattacharya, Homi Kharas, Mark Plant and Annalisa Prizzon (n 99) 7.

107 In 2019, 41 billion USD in climate finance were directed at developing and middle-income states, see *Joint Report on Multilateral Development Banks: Climate Finance* (2019) <https://www.ebrd.com/news/2020/mdbs-climate-finance-in-low-and-middleincome-countries-in-2019-reaches-us-415-billion.html>.

108 Joe Thwaites 'The good, the bad and the urgent: MDB climate finance in 2019' (18 August 2020) *WRI blog* <https://www.ebrd.com/news/2020/mdbs-climate-finance-in-low-and-middleincome-countries-in-2019-reaches-us-415-billion.html>. As discussed below, according E3G's evaluation tool the major MDBs are 'Paris aligned' with respect to the promotion of green finance, i.e. meeting the objectives of the Paris Agreement on this matter. See in Sonia Dunlop, James Hawkins, Zacharias Malik and Johannes Schroeten 'E3G Public bank climate tracker matrix' *E3G* <https://www.e3g.org/matrix/>.

109 MDBs 'boasted' in this respect that 'for every 1 dollar invested directly by MDBs in private sector operations, some 2–5 dollars are mobilized in additional private investment' *From Billions to Trillions: MDB Contribution to Financing for Development* (2015) <http://pubdocs.worldbank.org/en/69291436554303071/dfi-idea-action-booklet.pdf> 2.

110 Samantha Attridge and Lars Engen (n 199) 11; Thwaites (n 108).

111 Thwaites (n 108).

Other barriers with respect to the operation of development banks are discussed in the literature. In this respect, the OECD Secretariat mentions restrictions by shareholder countries, which affect the level of ambition that MDBs display in their policies.[112] Attridge and Engen also add MDBs' lack of appetite for risky investments, making investment in low-income countries less likely.[113] Studart and Gallagher further address the fact that development assistance often fails to 'spur emerging market ownership over the process, as the MDBs often dictate the terms and hold on to the expertise and technology'.[114] These authors further comment on the slow speed at which ODA's administrative funds often work ('World Bank loans for instance take 14–16 months to deliver'[115]) as well as the system's inadequacy 'in fostering local pipelines for projects and in creating local instruments for risk-sharing and the de-risking needed to crowd-in private capital'.[116]

Most of these issues demonstrate the importance of this study: on their own, these HCMs are certainly useful but somewhat inefficient (or at the least not as efficient as they could have been). As stated elsewhere in this study, the barriers that stand in the way of increasing private investment in the RE sector in the Global South are wide and varied: these barriers cannot be solved with the mere injections of funds, and other legal tools – from other areas of law – will have to be considered. The imperative to address a wider set of investors' needs implies a more prominent role for policy coherence, notably by exploring other international law instruments that will be linked to ODA funds and will improve their impact.

A Development banks and Policy Coherence for Development

The need to improve levels of policy coherence and align the operation of development institutions with climate laws and objectives, and even more specifically with the need to scale up investment in the RE sector, has been widely recognised.[117] To date, however, most of the discussion on the operation of development institutions and policy coherence has addressed

112 OECD, The World Bank, UN Environment (n 103) 102.
113 Samantha Attridge and Lars Engen (n 199) 12.
114 Rogerio Studart and Kevin Gallagher 'Guaranteeing sustainable infrastructure' (2018) 155 *International Economics* 84.
115 Studart and Gallagher (n 114) 90.
116 Studart and Gallagher (n 114) 90.
117 OECD, The World Bank, UN Environment (n 103) 100. The need to align development policies with climate-related objectives was recognised also by several national development agencies, a notable example being the French Development Agency, now committed to become the first "100% Paris Agreement" development-finance institution: "the first to systematically ensure that all project funding, in every country, finances resilient low-carbon development as compliant with the Paris Agreement". Agence Francaise de Developpment, *Towards a World in Common: AFD Group 2018-2022 Strategy* (2018) <https://www.afd.fr/sites/afd/files/2018-09-04-05-09/afd-group-strategy-2018-2022.pdf> 11.

internal coherence, notably within the world of development banks, at most also including links with other finance institutions.[118] This discussion is important, albeit limited in scope, as it does not review potential linkages with other HCMs from other regimes.

An important step in the right direction took place in 2017, when a group of nine development institutions (including the major MDBs and the World Bank) issued a Joint Statement in which they committed to align their operations with the objectives of the Paris Agreement, and to 'support low-emissions and climate-resilient development pathways through our interventions'.[119] The aspiration to align MDBs' activities with the objectives of a different international regime can certainly be regarded as indicative of a PCD-oriented approach. Indeed, this statement addresses the possibility of establishing 'collaborative partnerships with other institutions and private sector actors' as well as explicitly valuing 'the dialogue with the climate and finance communities'.[120] Moreover, this Joint Statement might carry more weight than intended: as discussed elsewhere in this study, the Paris Agreement itself addresses not only climate change, but also wider social objectives: Professor John Knox even referred to it as no less than a 'human rights treaty'.[121] While the Statement does not address human rights (possibly the signing parties did not intend to address this issue[122]), it could be that this 'implied' mandate will lead officials from these institutions to explore this area as well.

An analysis performed by the think tank E3G concluded that MDBs have yet to complete the alignment of their operations with the Paris Agreement, and in many areas progress is still required.[123] Notable shortcomings have been identified *inter alia* regarding the exclusion of fossil fuels, and, importantly for this study, with respect to access to energy.[124] With respect to

118 Amar Bhattacharya, Homi Kharas, Mark Plant and Annalisa Prizzon (n 99); Eugenia Heldt and Henning Schmidtke 'Explaining coherence in international regime complexes: How the World Bank shapes the field of multilateral development finance' (2019) 26(6) *Review of International Political Economy* 1160.

119 *The MDBs' Alignment Approach to the Objectives of the Paris Agreement: Working Together to Catalyse Low-Emissions and Climate-Resilient Development* (n 100).

120 *The MDBs' Alignment Approach to the Objectives of the Paris Agreement: Working Together to Catalyse Low-Emissions and Climate-Resilient Development* (n 100).

121 John Knox 'The Paris Agreement as a human rights treaty', in Dapo Akande, Jaakko Kuosmanen, Helen McDermott and Dominic Roser (eds.) *Human Rights and 21st Century Challenges* (OUP 2020) 323.

122 The OECD, for example, explains that 'Development co-operation that is Paris-aligned supports the three core objectives of the Paris Agreement on climate change mitigation, adaptation and finance flow consistency'. In other words, this understanding excludes the wider objectives of the Paris Agreement, including human rights. See OECD, *Aligning Development Co-operation and Climate Action: The Only Way Forward* (OECD 2019) 20.

123 Sonia Dunlop et al. (n 108).

124 Sonia Dunlop et al. (n 108).

'Promotion of Green Finance', however, most MDBs received a 'passing mark'; their policies deemed as 'Paris Aligned'.[125]

Another noticeable indication of MDBs' intentions to follow the 2017 Joint Statement is found in the European Investment Bank's (EIB) 2019 Energy Landing Policy.[126] In this policy, the EIB announced a shift in its investment policy towards an alignment with the objectives of the Paris Agreement. As part of this new policy, the EIB committed to several steps, most notably the phasing out of its support for fossil fuel-based energy projects.[127] Other important steps include a commitment to investment in energy efficiency, an increased support for RE projects[128] and, interestingly, 'support [for] the competitiveness of EU industry in low carbon technology'.[129] The EIB's policy signifies a positive step towards the alignment of climate-related objectives with public investment policies. However, it is hard to view this policy through the lens of PCD, as its focus seems mostly on the EU while development considerations have not been addressed. The EIB invests heavily in the Global South,[130] and it is unfortunate that this aspect has been omitted from its lending policy. Furthermore, it is also questionable whether an EU-led climate-investment policy that ignores the Global South could be regarded as 'aligned' with the Paris Agreement, or even as compatible with this agreement's core principles.[131]

Another important step taken by the EIB is the announcement of its transformation from 'an EU Bank supporting climate' into 'The EU Climate Bank'. This declaration was accompanied with the publication of the 'Climate Bank Roadmap' document, which explains how the EIB intends to deliver on this commitment.[132] Unlike its 'Energy Lending Policy', the EIB's Climate Bank Roadmap is far more development oriented.[133] The Roadmap includes important commitments, notably to increase the EIB's

125 Sonia Dunlop et al. (n 108).

126 European Investment Bank, *EIB Energy Lending Policy: Supporting the Energy Transformation* (EIB 2019) <https://www.eib.org/attachments/strategies/eib_energy_lending_policy_en.pdf>.

127 European Investment Bank, *EIB Energy Lending Policy: Supporting the Energy Transformation* (n 126) 14.

128 The EIB commits to extend it support for certain RE projects to a maximum of 75% of the project's cost, European Investment Bank, *EIB Energy Lending Policy: Supporting the Energy Transformation* (n 126) 21.

129 The EIB confirms that such support will take the form of subsidies for early stage technologies, see European Investment Bank, *EIB Energy Lending Policy: Supporting the Energy Transformation* (n 126) 23.

130 EIB 'The EIB and development' <https://www.eib.org/en/about/priorities/development/index.htm>.

131 Notably its commitment to the principle of common but differentiated responsibilities.

132 EIB, *Climate Bank Roadmap 2021-2025* (EIB 2020) <https://www.eib.org/attachments/strategies/eib_group_climate_bank_roadmap_en.pdf>.

133 The Roadmap states *inter alia*: 'The opportunities for the EIB to strengthen its support towards EU external action through development financing are also strong, to ensure

support for 'climate action and environmental sustainability' to a level that will exceed 50% of its lending by 2025, aspiring to leverage €1 trillion of investment.[134] The EIB further reiterates its commitment to aligning its activity with the Paris Agreement, and adds a commitment that its activities will 'do no significant harm to the low-carbon and climate resilient goals of the Agreement'.[135]

While certainly a step in the right direction, the EIB's Roadmap was criticised by civil society organisations, notably due to alleged 'loopholes', allowing the EIB to continue financing fossil fuels-based projects as well as other investments in airports and roads which are 'the textbook definition of projects that a climate bank should stop supporting'.[136]

B Development institutions and safeguards

Before concluding the (somewhat brief) discussion on development institutions, some attention should be given to the safeguards that development institutions rely on. As stated elsewhere, the focus of this study is on HCMs, and not so much on safeguards. Reliance on safeguards can be seen as 'first-generation' integration (i.e. attempts to ensure that investors do no harm), and as such distinguished from the more active efforts *to promote* environmentally sound investment policies, which are the focus of this study. At the same time, it is important to at least mention these safeguards, especially in those areas that are less known to international law scholars.[137]

A review of development institutions' activity reveals a clear effort to impose safeguards on investors' operations. Leading examples include the International Finance Corporation (IFC) Performance Standards[138] that

---

that EU external action, climate action and development cooperation is more than the sum of its parts'. EIB, *Climate Bank Roadmap 2021-2025* (EIB 2020) (n 132) 5.

134 EIB, *Climate Bank Roadmap 2021-2025* (n 132) 1.

135 EIB, *Climate Bank Roadmap 2021-2025* (n 132) vi.

136 Xavier Sol 'The EIB has further to go to become the "EU Climate Bank"' (July 7 2020) *Climate Home News* <https://www.climatechangenews.com/2020/07/07/eib-go-become-eu-climate-bank/>; 'Dreams of an EU Climate Bank go up in smoke' (11 November 2020) *Counter Balance* <https://counter-balance.org/news/dreams-of-an-eu-climate-bank-go-up-in-smoke>; WWF 'EIB as EU 'climate bank' must require pollutes to change' (10 November 2020) *WWF* <https://www.wwf.eu/?uNewsID=1085366>.

137 The literature on safeguards is abundant mostly in areas such as international investment law and international trade law. See for example Avidan Kent and Markus Gehring 'Investment law and the environment: Evolving international practice and norms' in Erika Techera, Jade Lindley, Karen Scott and Anastasia Telesetsky (eds.) *Routledge Handbook of International Environmental Law* (2nd ed. Routledge 2021); Steve Charnovitz 'Exploring environmental exceptions in GATT Article XX' (1991) 25 *Journal of World Trade* 37.

138 See review of the standards and related documents on the IFC's website <https://www.ifc.org/wps/wcm/connect/Topics_Ext_Content/IFC_External_Corporate_Site/Sustainability-At-IFC/Policies-Standards/Performance-Standards>.

many other institutions indeed followed,[139] as well as the World Bank's Environmental and Social Framework.[140] While the author has not reviewed the entire universe of development institutions, it seems likely that the vast majority of these institutions (including all leading MDBs) have adopted somewhat similar mechanisms, most of which aim to mitigate investors' adverse effects on areas such as environmental protection and human rights.

A cursory review of these standards and frameworks reveals very little attention to RE, although wider references to climate change do exist. As identified by Morgado and Taskin,[141] some of these safeguards address greenhouse gases (GHG) emissions, albeit mostly in a 'defensive' manner, i.e. as an instruction to avoid GHG pollution,[142] rather than actively encouraging new investment in the RE sector. Interesting examples include the rather soft requirements to 'consider alternatives and implement technically and financially feasible and cost-effective options to reduce project-related GHG emissions',[143] and (in the case of big polluters) to 'quantify direct emissions from the facilities'.[144]

As stated by Morgado and Taskin, the implementation of development institutions' safeguards leaves much to be desired.[145] Notable challenges include the lack of completing domestic regulatory framework (e.g. environmental and human rights laws), as well as deficient enforcement.[146] These authors further mention problems with the design and monitoring of relevant plans, due to insufficient expertise and human resources.[147]

Also importantly, while naturally vital, the focus on 'no harm' seems somewhat outdated. As advocated throughout this study, this focus is insufficient as it does little to promote investment. Indeed, certain states are currently advocating for a different, more active approach with respect to development-related investment and its alignment with climate-related objectives,[148] albeit with little result.

An important (even if modest) step towards a shift from a strictly 'no harm' approach, to more active support of sustainable investment, is found

139 See review in Naeeda Crishna Morgado and Ozlem Taskin, *Managing Environmental Risks in Development Banks and Development Finance Institutions – What Role for Donor Shareholders?* (OECD 2019) 17.

140 *The World Bank Environmental and Social Framework* (World Bank 2017) <http://pubdocs.worldbank.org/en/837721522762050108/Environmental-and-Social-Framework.pdf>.

141 Naeeda Crishna Morgado and Ozlem Taskin (n 139) 39.

142 See for example the IFC Standard 1, which requires as part of risk assessments to address also GHG emissions, or Standard 3, which requires the reduction of project-related GHG.

143 International Finance Corporation, *Guidance Note 3: Resource Efficiency and Pollution Prevention* (IFC 2012) 5.

144 International Finance Corporation, *Guidance Note 3* (n 143) 5.

145 Naeeda Crishna Morgado and Ozlem Taskin (n 139) 21.

146 Naeeda Crishna Morgado and Ozlem Taskin (n 139) 21.

147 Naeeda Crishna Morgado and Ozlem Taskin (n 139) 21.

148 Naeeda Crishna Morgado and Ozlem Taskin (n 139) 24–25.

in the 2016 G7 Ise-Shima Principles for Promoting Quality Infrastructure Investment,[149] and their 2019 'successor' G20 Principles for Quality Infrastructure Investment.[150] These principles call on a list of stakeholders (including MDBs) to align infrastructure investment with several key principles. Notably, Principle 4 of the Ise-Shima Principles dictates the alignment of development policies with climate change strategies, including the 'promotion of private investment and attraction of new industries'.[151] Principle 5 adds in this respect a call for 'effective resource mobilization [...] including through MDBs'.[152]

While the Principles' language on sustainability, integration and investment promotion is indeed important, their lack of specific instruction on RE-related investment has not gone unnoticed. As observed by Aizawa: 'The drafters, however, failed to deliver on climate change, which is truly unfortunate because this renders the overall Principles less balanced than they initially appear, and for the climate advocates, almost useless'.[153] At the same time, defining these Principles as 'almost useless' seems an exaggeration: these Principles signify an important shift in approach towards the global efforts to integrate development with foreign investment, and as such they are also likely to affect more specific action with respect to climate-related investment. Indeed, the same author commented that 'the Principles are noteworthy for offering an authoritative narrative on what should matter in infrastructure development and investment that is yet to come'.[154]

IV Development treaties

The final instruments to be discussed in this chapter are development agreements. As reviewed below, these agreements may take the shape of 'proper' treaties between states, or, alternatively, the form of more limited and targeted arrangements between states' national agencies and institutions. In both cases, these instruments fall within the definition of international law HCMs.

The shape of development agreements varies to a great extent, as does the variety of instruments used by these agreements. Although often based on economic tools such as trade and investment, development agreements are distinguished from trade and investment treaties in terms of language,

149 *G7 Ise-Shima Principles for Promoting Quality Infrastructure Investment* (2016) <https://www.mofa.go.jp/files/000160272.pdf>.

150 *G20 Principles for Quality Infrastructure Investment* (2019) <https://www.mof.go.jp/english/international_policy/convention/g20/annex6_1.pdf>.

151 Principle 4, *Ise-Shima Principles for Promoting Quality Infrastructure Investment* (n 149).

152 Principle 5 *Ise-Shima Principles for Promoting Quality Infrastructure Investment* (n 149).

153 Motoko Aizawa 'Sustainable development through quality infrastructure: Emerging focus on quality over quality' (2019) 1(2) *Journal of Mega Infrastructure & Sustainable Development* 171, 182.

154 Motoko Aizawa (n 153), 181.

objectives and types of commitments. Development agreements are typically concluded between developed and developing parties. Their primary objective is a-symmetrical in nature: to achieve development *in the developing party/ies* to the agreement. Very often, the parties intend to achieve this objective through the implementation of a variety of instruments, notably from the world of economic law. As such, alongside development-oriented language (including commitments to provide ODA[155]), these agreements often include legal instruments such as rules on the trade in goods, trade facilitation, technical barriers to trade, SPS, trade in services, e-commerce, intellectual property and more. The inclusion of these 'commercial' aspects under the title of 'development agreements' forces negotiators to consider a wider range of interests, instruments and safeguards. These instruments are, in a way, tailor-made HCMs, adjusted to fit the commercial needs and capabilities of the developed party, with the development needs of the developing party. As such, the relevance of policy coherence for the design of these agreements is clear.

An example that demonstrates the hybrid nature of development agreements is found in the EU Commission's description of the CARIFORUM-EC Economic Partnership Agreement (EPA)[156]:

> The EPA is the trade and development partnership signed in 2008 by the 15 states of CARIFORUM and the EU's 28 countries. Our goals are simple: to make it easier for people and businesses from our two regions *to invest in and trade* with each other, and to *spur development* across the Caribbean.

While the EU's model of development and cooperation treaties is probably the most well-known and oft-discussed in the literature,[157] the reader should be aware that other models exist as well.[158] Interestingly, the most recent EPA model *could have been* the UK's newly concluded, post-Brexit EPAs. A cursory review of these agreements, however, reveals that these are mostly 'copy-pasted' from the EU's own agreements, and as such, it is difficult to address them as 'new'.

On the other hand, and as reviewed below, much of the commitments found in the EU's EPAs, and now transplanted into the UK's treaties, are

155 See for example Article 36(5) of Interim Agreement establishing a framework for an Economic Partnership Agreement between the Eastern and Southern Africa States, on the one part, and the European Community and its Member States, on the other part.

156 Emphasis added. EU Commission, *Factsheet: How the EU Is Putting the EPA Into Practice* (2018) https://trade.ec.europa.eu/doclib/docs/2012/april/tradoc_149286.pdf>.

157 Avidan Kent 'Policy coherence and the promotion of foreign direct investment in the renewable energy sector: Lessons from Europe' in Lisa Sachs, Lise Johnson and Jesse Coleman (eds.) *2017 Yearbook on International Investment Law and Policy* (OUP 2018) 22.

158 See for example Australia and Papua New Guinea's Treaty on Cooperation and Development (2000).

'open' in nature (not to say plainly vague). The implementation of these commitments will therefore depend on the UK's political will, which will be independent and different from that of the EU. This distinction creates a wide window of possibilities: it could be that the UK will fill these commitments with meaningful content and support the promotion of RE-related investment in the Global South. At the same time, these commitments are vague enough to allow the UK to simply dismiss them as empty political statements, leading to nothing at all. In any event, this window of possibilities is interesting to follow, and, accordingly, this chapter will also pay close attention to the UK's legal obligation (even if most of those were not originally 'authored' by the UK).

A Examples

There are numerous examples of development agreements. Due to limitations of scope and space, the author will only review a handful. Special attention will be given to RE-related language that could be seen as supportive of the efforts to promote FDI in the RE sector in the Global South.

One notable example of a development agreement is the Cotonou Agreement between the EU and the African, Caribbean and Pacific Group of States (ACP).[159] As stated above, this agreement attempts to integrate development needs with economic law HCMs. Indeed, this agreement mentions a wide variety of objectives, *inter alia* the eradication of poverty, sustainable development and the integration of ACP countries into the global economy. With respect to policy coherence, the agreement states[160]:

> These objectives and the Parties' international commitments, including the Millennium Development Goals, shall inform all development strategies and shall be tackled through an integrated approach taking account at the same time of the political, economic, social, cultural and environmental aspects of development. The partnership shall provide a coherent support framework for the development strategies adopted by each ACP State.

Importantly for this study, the Cotonou Agreement includes a variety of instructions on trade and investment, albeit nothing on investment protection. With respect to the promotion of investment in the RE sector, this agreement includes several, fairly vague references to the importance of

159 A modernised version of this agreement will enter into force in 2021. At the time of writing, the EU Commission did not agree to share a copy of the concluded agreement.

160 Article 1, Cotonou Agreement between the EU and the African, Caribbean and Pacific Group of States (ACP).

renewables,[161] none of which could be described as meaningful. However, the Cotonou Agreement is being re-negotiated at the time of writing, and a new, modernised version of this agreement is expected to enter into force in 2021.[162] While the text of the new agreement has not been made available to the author,[163] an announcement released by the EU Commission leaves some room for optimism. This announcement confirms that 'environmental sustainability and climate change' have been defined as 'key priority' areas, and that the agreement will be a 'powerful tool' to advance the SDGs.[164] Also importantly, 'a renewed focus on climate change' was announced, including a specific reference to the Paris Agreement.[165] Whatever shape these statements take in terms of legal obligations, and whether they will translate into real, meaningful action remains to be seen.

Another example is found in an EPA concluded between the UK and Eastern and Southern Africa (ESA) States, which, as stated above, was mostly copy-pasted from the EU-ESA EPA. The preamble to this agreement confirms that this agreement should 'serve as an instrument for development' and lead to increased trade and investment.[166] This agreement sets as a treaty objective[167]:

> contributing to the reduction and eventual eradication of poverty through the establishment of a *strengthened and strategic trade* and development partnership consistent with the *objective of sustainable development and the Sustainable Development Goals (SDGs).*

This agreement is accordingly focussed on international trade and the reduction of trade barriers. With respect to investment in the RE sector, Article 37 of the UK-ESA EPA also targets as an important area for cooperation 'trade related infrastructure, including transport, energy and water'. Article 46 of this agreement (copied from Article 47 of the EU-ESA EPA) further emphasises that 'The Parties recognise the importance of cooperation in the energy sector as a vehicle for supporting the ESA economies' competitiveness at the regional and global levels'. This Article also sets objectives with respect to energy-related cooperation, including

161 See for example Article 32, Cotonou Agreement between the EU and the African, Caribbean and Pacific Group of States (ACP).
162 European Commission, Questions and Answers: Reaching a political deal at chief negotiators' level on the EU's new Partnership Agreement with members of the OACPS' (3 December 2020) <https://ec.europa.eu/commission/presscorner/detail/en/qanda_20_2303>.
163 The European Commission refused to provide a copy.
164 European Commission (n 162).
165 European Commission (n 162).
166 Preamble, Agreement establishing an Economic Partnership Agreement between the Eastern and Southern Africa States, on the one part, and the United Kingdom of Great Britain and Northern Ireland, on the other part.
167 Art 1, UK-ESA EPA (n 166).

the objective to 'improve the access of ESA States to modern, efficient, reliable, diversified and sustainable and renewable sources of clean energy at competitive prices'.[168]

This Article is important: the language included in this legal provision closely follows the imperative set in SDG 7, and could be seen as laying the foundations for a more concrete mechanism that will deliver on this front. While the language included in this provision is, admittedly, vague, Article 51 provides it with an important, additional level: the UK commits to supporting the implementation of the EPA,

> *through appropriate and effective UK funding mechanisms* in consultation with ESA partners to *contribute to implementing the programmes and projects to be developed under the areas of cooperation identified in this Agreement* and relevant chapters and under the detailed Development Matrix.[169]

Also importantly, the Parties to the UK-ESA EPA commit to creating a Joint Development Committee that will monitor the implementation of commitments regarding development cooperation, including those discussed above.[170] While the obligation to create a Joint Development Committee is also copy-pasted from the EU-ESA EPA, it seems that such a joint-institution (possibly under a different title) will indeed be established.[171] The prospective establishment of such a committee will certainly support the implementation of these commitments.

Important as they are, these commitments are, for the time being, vague. It will therefore be interesting to see what kind of content the UK will read into these provisions; how many resources will be dedicated to their implementation, and in what way such resources will be used.

Other interesting examples are found in the UK-CARIFORUM EPA (which was mostly copy-pasted from the EU-CARIFORUM EPA).[172] This agreement is also committed to a holistic, PCD-oriented approach, mentioning trade policy and private sector investments as tools for the attainment of

168 Art 46, UK-ESA EPA (n 166).

169 Art 51 UK-ESA EPA (n 166).

170 Art 51(2) UK-ESA EPA (n 166).

171 UK Department of International Trade 'Continuing the United Kingdom's trade relationship with the CARIFORUM States' (2019) <https://assets.publishing.service.gov.uk/government/uploads/system/uploads/attachment_data/file/803587/CARIFORUM-UK-EPA.pdf> 19.

172 Economic Partnership Agreement between the CARIFORUM States, of the one part, and the United Kingdom of Great Britain and Northern Ireland, of the other <https://www.gov.uk/government/publications/ms-no182019-economic-partnership-agreement-between-the-cariforum-states-of-the-one-part-and-the-united-kingdom-of-great-britain-and-northern-ire>.

the SDGs.[173] Chapter 5 of Title IV of this agreement (titled 'social aspects') reaffirms commitments made under other multilateral agreements (notably labour-related) and highlights 'the value of greater policy coherence between trade policies, on the one hand, and employment and social policies on the other'.[174]

More specifically and with respect to RE, this agreement includes a specific, yet, somewhat loose ('resolved to made efforts') commitment[175]:

> The Parties and the Signatory CARIFORUM States are resolved to make efforts to facilitate trade in goods and services which the Parties consider to be beneficial to the environment. Such products may include environmental technologies, renewable- and energy-efficient goods and services and eco labelled goods.

More concrete language appears in Article 138, which is entirely focussed on the cooperation between the Parties on eco-innovation and RE. According to this Article, the Parties commit to cooperate, 'including by facilitating support', in a variety of projects, *inter alia* 'related to energy efficiency and renewable energy'.[176] Other related areas for envisioned cooperation include research, technical support, exchange of information/know-how, and pilot and demonstration projects.

On the face of it, the commitment made in Article 138 appears promising. On the other hand, it is not entirely clear what the vague term 'facilitating support' really means. The reader should also be aware that this commitment is made under Chapter 2 – 'innovation and intellectual property'. In light of this chapter's objectives, it could be that this Article's scope will be limited to new technologies, while well-established RE technologies will not be supported.

Once again, the fact that the UK is following the EU's footsteps so closely is a useful point of departure. Whether intentional or not,[177] the UK's development agreements explicitly address investment in the RE sector, as well as in a line of other competing areas, such as the potential liberalisation of trade in Environmental Goods and Services (see above) and the protection of intellectual property rights.[178] This approach is certainly in line with the notion of PCD. At the same, to date, much of this language is still too vague

173 While the EU-CARIFORUM EPA refers to the older Millennium Development Goals, Art 1 of the CARIFORUM-UK EPA (n 172) is already referring to the 'updated' SDGs.

174 Art 191 of the CARIFORUM-UK EPA (n 172).

175 Art 183(5) of the CARIFORUM-UK EPA (n 172).

176 Art 138(5) of the CARIFORUM-UK EPA (n 172).

177 The UK more or less bulk 'copy-pasted' thousands of pages from many existing EU agreements into its own new agreements. It is therefore questionable whether the UK policy-makers indeed scrutinised each and every legal provision in these agreements.

178 Chapter 2, Section 2, CARIFORUM-UK EPA (n 172).

and the exact meaning that the parties will, eventually, pour into these commitments is not yet clear.

Lastly, as stated above, development agreements may also take the shape of a targeted arrangement between national agencies.[179] An interesting example is a Memorandum of Understanding (MOU) signed between the investment promotion and development agencies of the US (OPIC), Australia (EFIC) and Japan (JBIC).[180] This MOU targets investment in the Indo-Pacific region, and it specifically mentions the energy sector as included within its scope of cooperation.[181] This agreement is therefore of relevance for this study.

This MOU is admittedly very brief and lacking in content. It does, however, specify that the three parties will share information and expertise on investment locations and sectors and jointly finance projects[182] – all activities that fall within the definition of HCMs.[183] Interestingly, Art 3 of this MOU states that the parties will 'engage in other forms of cooperation' as appropriate, creating a door for other HCMs, such as, potentially, the use of Policy Risk Insurances (PRI).

B Development treaties and PCD: the missing bits?

As stated above, in comparison with other international law instruments, development treaties indeed present a relatively high level of policy coherence. The reviewed examples include provisions on ODA, alongside (mostly) trade rules and development-related safeguards and objectives. These agreements rely on a variety of HCMs, *inter alia* technical assistance and information exchange (second type of HCMs[184]), rules on trade liberalisation (sixth type of HCMs) and, importantly, financial incentives (fourth type of HCMs).

There is no doubt that certain elements from these agreements should be used as a model, to be followed in other areas of international law.

179 See for agreement a Regional Development Cooperation Agreement between the U.S. Agency for International Development (USAID) and the Association of Southeast Asian Nations (ASEAN), concluded in 2020 <https://kh.usembassy.gov/usaid-and-asean-sign-inaugural-regional-development-cooperation-agreement/#:~:text=%E2%80%9CThis%20new%20five%2Dyear%20agreement,in%20ASEAN%2C%E2%80%9D%20expressed%20USAID%20Principal>; Agreement between the Norwegian Agency for Development Cooperation (Norad) and the Government of the Republic of Iraq in connection to development cooperation <https://www.norad.no/contentassets/d2e63738e9c24af4a9faf1c04ccdf87d/programme-agreement.pdf?id=22260>.

180 Memorandum of Understanding on Cooperation in the Indo-Pacific Region (2018) <https://www.opic.gov/sites/default/files/files/181112_Trilateral_MOU-signed.pdf>.

181 See Art 2 of this MOU (n 180).

182 See Art 3 of this MOU (n 180).

183 These activities are falling within the UNCTAD typology's second type of HCMs. UNCTAD, *Home Country Measures* (UNCTAD 2001) 9.

184 UNCTAD (n 183) 9.

Notably, the holistic approach expressed in these agreements is useful, and so is the reliance on a *variety* of instruments. The different instruments are seen as *completing* one another's utility and improving their effectiveness. This approach is not surprising: as stated in the introduction to this chapter, the international community is anxious to improve the effectiveness of development assistance, and the recourse to PCD in this respect is understandable.

Other desirable elements include the successful mainstreaming of development considerations throughout these agreements. The importance of development-related considerations re-appears throughout these agreements. The reliance on development as a leading guideline does not end with preambular declarations; it is evident in the detailed design of specific instructions. This is a lesson in mainstreaming, which treaty negotiators would do well to learn from.

At the same time, these agreements are far from perfect. In the context of this study, it is clear that despite the obvious impact that RE-related investment could have on development, this area has hardly been explored.

To begin with, many of these agreements address the RE sector only very vaguely, and often in a declaratory manner, with no accompanying concrete rules. Many of the reviewed provisions fall under the first type of HCMs – policy positions – 'positive in tone but vague in specific commitments'.[185] There is no doubt that even if vague, this type of language carries some weight: ODA funds that are promised through these agreements indeed find their way to green investments, including in the RE sector. The lack of concrete language, however, is problematic in terms of certainty and long-term expectations.

Another shortcoming of existing development agreements is related to the fairly ambiguous language that is included on investment rules. On the one hand, statements concerning the importance of FDI are common. For example, the EU-ESA EPA states the following as a treaty objective[186]:

> establishing and implementing an effective, predictable and transparent regional regulatory framework for trade and investment in the ESA region, thus supporting the conditions for increasing investment and private sector initiative and enhancing supply capacity, competitiveness and economic growth

On the other hand, beyond repeatedly stressing the importance of investment promotion and *protection*, most of these agreements do not include much. Notably, a chapter on investment protection is routinely absent. The omission of investment protection rules is the result of the ever so

185 UNCTAD (n 183) 8.
186 Art 2 EU-ESA EPA (n 155). See also Article 40 of this agreement.

popular criticism, repeatedly made against the international investment regime. As explained elsewhere in this study,[187] much of this criticism is exaggerated, blunt and even misguided: while the rules on investment protection should, without a doubt, be designed with great care and attention,[188] avoiding them altogether is tantamount to throwing the baby out with the bath water. Importantly, the rules on investment protection are important where investment conditions are difficult, especially for investments that require high upfront, sunk costs, which is the case in the RE sector.[189] If indeed promoting FDI is a priority, as is the effectiveness of development assistance, such rules should have not been avoided. Even some of those who have criticised the investment regime in the past have agreed that much more attention should be given to investment rules in development agreements. The need to add a designated investment chapter to the Cotonou Agreement was stressed as early as 2003 by Konrad von Moltke[190]:

> A Cotonou Investment Agreement is needed to amplify investment provisions of the CPA to identify the full range of issues that need to be addressed to ensure that a Cotonou Investment Agreement promotes the objectives of the CPA.

Another area that does not receive sufficient attention in development agreements is insurance. The importance of insurance mechanisms has been clarified elsewhere in this study[191]: these tools can significantly reduce the cost of investment and address investment barriers. A cursory review of EU and UK Preferential Trade Arrangements reveals that very little attention has been given to the use of insurance mechanisms as tools that could promote investment in the Global South. A unique and fairly limited example is found in the EPAs concluded by the EU, and the UK, with ESA Countries. Insurance mechanisms have been identified in these agreements as 'areas of cooperation' with respect to private sector development. The parties accordingly committed to[192]:

187 See more on this topic in Chapter 3.

188 For further review see Avidan Kent and Markus Gehring 'Investment law and the environment: Evolving international practice and norms' in Erika Techera, Jane Lindley, Karen Scott and Anastasia Teletsky (eds.) *Routledge Handbook of International Environmental Law* (Routledge 2020), Chapter 27.

189 Lisbeth Colen, Damiaan Persyn and Andrea Guariso 'Bilateral investment treaties and FDI: Does the sector matter?' (2016) 83 *World Development* 193.

190 Konrad von Moltke *A Cotonou Investment Agreement: Report for the Commonwealth Secretariat* (Commonwealth Secretariat 2003) <https://www.iisd.org/system/files/publications/investment_cotonou.pdf>.

191 See more on this topic in Chapter 4.

192 Art 39(2)(h)-(i), UK-ESA EPA (n 172); Art 40(2)(h)-(i) EU-ESA EPA (n 155).

ensure the increasing availability and use of risk insurance as a risk-mitigating mechanism in order to boost investor confidence in the ESA States;

and (i) offer guarantees and assist with guarantee funds covering risks for qualified investment.

As stated elsewhere in this study, authors such as Paul Collier have envisioned a more targeted use of ODA funds, notably for covering the cost of insurance premiums, *inter alia* for infrastructure projects in Africa.[193] Once again, the UK has an opportunity in this area: if taken seriously, this declaratory language could lead to meaningful and effective action.

Lastly, the link with the environmental regime is limited as well. Admittedly, most of the efforts to achieve linkages between the development and the environment regimes are made under the UNFCCC, where efforts to design successor mechanisms for the Clean Development Mechanism (CDM) are made. While these multilateral efforts have been discussed elsewhere in this study,[194] one may ask why similar efforts are not made on a bilateral (or regional) basis as well. Just as nations are linking emission trading systems,[195] developed nations could consider bilateral/regional CDM-like mechanisms that will incorporate reductions in targeted developing nations into, for example, their own emission trading schemes.

V Conclusion

As reviewed in this chapter, there is little doubt that development-related HCMs could play, and already are playing, an important role in the efforts to increase the flow of investment towards the Global South. ODA funds could address a variety of investment barriers, ranging from providing finance for improved infrastructure, to capacity building, the reduction of investment costs,[196] supporting (or subsidising) high levels of returns[197] or the mitigation of certain risks.[198] If targeted and designed

193 Paul Collier 'Attracting international private finance for African infrastructure' (2014) 1 *Journal of African Trade* 37, 41.

194 Chapter 7 of this study.

195 Marissa Santikarn et al., *A Guide to Linking Emission Trading Systems* (ICAP 2018).

196 Nicholas Stern 'Understanding climate finance for the Paris summit in December 2015 in the context of financing for sustainable development for the Addis Ababa conference in July 2015' (2015) Policy paper, ESRC Centre for Climate Change Economics and Policy Grantham Research Institute on Climate Change and the Environment <https://www.lse.ac.uk/GranthamInstitute/wp-content/uploads/2015/03/Stern-policy-paper-March-20151.pdf> 5.

197 The IPCC 2014 report reviews the rates of return expected by investors in low-carbon sector, and summaries that '[m]any renewable energy projects, especially in developing countries where additional risk margins are added, are struggling to reach returns of this level to satisfy the expectations of financiers of equity and debt'. IPCC Sujata Gupta et al. (eds) 'Chapter 16: Cross-cutting Investment and Finance Issues' in *Climate Change 2014: Mitigation of Climate Change* (IPCC 2014) 1223.

198 Elizabeth Asiedu et al. 'Does foreign aid mitigate the adverse effect of expropriation risk on foreign investment? (2009) 78(2) *Journal of International Economics* 268.

properly, ODA could supplement and enhance other HCMs and improve their effectiveness.

At the same time, it is widely agreed today that existing ODA commitments will not be enough: these sums, *on their own*, are insufficient to support a genuine transition into a green economy and to achieve the objectives of the Paris Agreement or the SDGs. As reviewed in this chapter, policy-makers are aware of these deficiencies, and, accordingly, address ODA as a tool for *leveraging* private finance, hoping that effective leveraging will satisfy investment's needs. Sadly, these efforts, to date, have not been sufficient,[199] and new approaches will have to be explored.

Moreover, the above implies that on their own, mere ODA transfers do not address the myriad of investment barriers that stand in the way of foreign investment in the Global South. Indeed, when assessing the failure of ODA in leveraging private investment, Attridge and Engen commented[200]:

> Private commercial finance *will not flow freely to countries where the investment climate is challenging*, markets are not functioning and the risk-adjusted rate of return is uncompetitive. In fact, we estimate that 96.3% of private finance mobilised through blended finance flows to countries with a credit rating, which most LICs do not have. Blended finance may tip the balance, but it will not work if the economic fundamentals are not in place. *So, the push for blended finance should not eclipse the need for grants to boost local investment environments.*

In other words, PCD is not only recommended, but is in fact crucial for the effectiveness of ODA. Other strategies that will complete the transfers of ODA funds with rules on trade, investment, finance and insurance will improve ODA's value for money, address investment barriers and convince more investors to put their money where it is genuinely needed. As reviewed above, this insight is not novel on its own and in some areas – notably development agreements – states are incorporating complementary HCMs in order to improve investment conditions and achieve better results. As argued in this chapter, development agreements represent a useful model in this respect, even if an insufficient one.

199 The Overseas Development Institute study found that $1 invested by development institutions, mobilised only $0.37 in low-income countries, $1.06 in lower-middle-income countries and $0.65 in upper-middle income countries. Samantha Attridge and Lars Engen, *Blended Finance in the Poorest Countries* (Overseas Development Institute 2019) <https://www.odi.org/sites/odi.org.uk/files/resource-documents/12666.pdf> 11.

200 Emphasis added. Attridge and Engen (n 199) 11.

As reviewed in this chapter, policy-makers have acknowledged the need to improve policy coherence, and have welcomed the role that PCD could play in improving the effectiveness of development assistance. As also reviewed, these efforts, to date, have been limited and the search for coherence has not extended far enough. Policy-makers will do well to explore linkages with a variety of international law HCMs, including those discussed in this study.

7 International environmental law (and human right law, too)

I Introduction

It is widely recognised that environmental laws and policies should play a central role in the process of global development. As discussed in the first part of this study, Sustainable Development Goal (SDG) 13 explicitly calls for action on climate change as part of the wider UN's development agenda. SDG 7 clarifies in this respect, that climate action should be integrated into development policies through a 'significant' increase in the share of renewables 'in the global energy mix'.[1] While the term 'renewable energy' is not explicitly mentioned in the UNFCCC itself (or in the more recent Paris Agreement), it is clear that the promotion of RE is an inherent objective of the climate regime. Indeed, some of the UNFCCC's key mechanisms and funds are geared towards the attainment of this goal.

The following section will review certain specific environmental law instruments. This section will identify relevant international law Home Country Measures (HCMs) and will evaluate the state of Policy Coherence for Development (PCD) in this area, if such is to be found. More attention will be given to interaction with areas that have not been covered so far in this book, notably human rights law.

II International environmental law and the promotion of RE-related investment

A The Kyoto model

As stated above, the UNFCCC does not explicitly mention the term 'renewable energies'. It is nevertheless evident that very early on in the process – during the negotiation of the Convention itself – the role of alternative energy sources was considered as part of the effort to stabilise greenhouse gases (GHG) levels in the atmosphere at a level that will be regarded as safe.

1 Target 7.2 of SDG 7. UNGA, Resolution adopted by the General Assembly on 25 September 2015, UN Doc A/RES/70/1.

DOI: 10.4324/9780429277207-8

A clear indication of the Convention's intentions in this context is found in its preamble, according to which the Parties are[2]:

> *Recognizing* that all countries, especially developing countries, need access to resources required to achieve sustainable social and economic development and that, in order for developing countries to progress towards that goal, their energy consumption will need to grow taking into account the possibilities for achieving greater energy efficiency and for controlling greenhouse gas emissions in general, *including through the application of new technologies* on terms which make such an application economically and socially beneficial.

In order to promote the adoption of these 'new technologies', the UNFCCC Member States adopted in 1997 the Kyoto Protocol (KP): a market-based legal mechanism that was regarded at the time as a novelty. The KP was geared towards the reduction of emissions' levels and its outcome was to be measured in two periods: the first ended in 2012 and the second in 2020.

Even if not yet officially terminated,[3] it is clear that the KP should be evaluated mostly in historical terms; the KP has not been extended beyond 2020 and is considered at the time of writing as mostly inoperative.[4] The KP is nevertheless an important legal mechanism, one with special relevance to the topic of this study and because of the many valuable lessons to be learned from it. Moreover, the model attempted by the KP could be refined and improved, as indeed is the intention of the Member States (see discussion on the Sustainable Development Mechanism, below).

The following part will address mostly the KP's flexible mechanisms as the operation of these mechanisms fits neatly within the definition of international law HCMs. Notably, the KP's Joint Implementation (JI) and Clean Development Mechanism (CDM) were both designed to encourage investment in renewable energy projects through the provision of financial incentives. The purpose of these flexible mechanisms was twofold: to allow developed nations the necessary flexibility in meeting their legally binding emission reduction targets, while supporting developing nations' efforts to reduce their own emission levels. The CDM (Art 12 KP) is more relevant

2 UNFCCC, preamble, emphasis added.

3 An official termination of the Kyoto Protocol cannot be easily achieved as it requires either the consent of all parties, the conclusion of later treaty on similar subject matter or a prolonged abandonment of this treaty, all of which are unlikely to occur, see Daniel Bodansky et al., *International Climate Change Law* (CUP 2017) 205–206. It is nevertheless clear that the Kyoto Protocol can hardly be regarded as a living legal tool post-2020 due to the lack of political interest/engagement.

4 See discussion on the future of the CDM post-2020 in Moritz Unger, Sandra Greiner and Nicole Krämer, *The CDM Legal Context Post-2020: Discussion Paper* (Climate Focus 2019) <https://www.climatefocus.com/sites/default/files/ClimateFocus_Atlas_2019_CDM_Post_2020.pdf>.

for the purpose of this study as it addresses RE-related investment in the Global South, and will therefore receive more attention in this study.

The mechanisms themselves are simple in nature, but complex in terms of technicalities and implications. In a nutshell, where investors from the developed world (Annex I countries) were able to reduce host-states' emissions through their investment (e.g. by building a RE facility), they were issued with Certified Emission Reductions (CERs) which could be later sold through carbon markets. In other words, the CDM offers an additional financial incentive for RE investment in the developing world, and therefore fits within the United Nations Conference on Trade and Development's (UNCTAD) fourth group of HCMs (financial and fiscal incentives).[5]

According to some, CDM projects 'have unequivocally contributed to improving the attractiveness of projects in the electricity sector'.[6] At the time of writing, there are 8,053 CDM projects (7,848 are registered[7]) in 111 non-Annex I countries.[8] According to the UNFCCC Secretariat, these projects reflect an investment of 303.8 billion USD, leading to a reduction of approximately 2 billion tonnes of CO_2 in the developing world.[9] According to the Executive Secretary of the UNFCCC, Patricia Espinosa[10]:

> The CDM exceeded everyone's expectations. [...] the CDM mobilized the private sector – essentially sending project developers off in search of low-cost climate change mitigation opportunities in developing countries. [...] The benefits go far beyond mitigation. [The financing that went into developing countries – including USD 200 million contributed to the Adaptation Fund through a 2 per cent levy on the issuance of CERs – has improved the lives of millions of people and contributed to nations' sustainable development goals. More than 1 million efficient cookstoves were installed, reducing emissions and improving health.

5 Elsewhere I have made the point that it is not entirely clear whether these mechanism are indeed HCMs, as the financial support itself (i.e. the entity that will pay for the emission units) is not necessarily the investor's home state. On the other hand, it could also be asserted that the investor's home state has 'paid' for its inclusion in Annex I by undertaking expensive binding emission cuts. Therefore, the financial contribution obtainable via the CDM is directly linked to the home state's economic sacrifice.

6 Adelino Ricardo J. Esparta and Karen M. Nagai 'Experiences and lessons of CDM in the energy sector' in Flavia Witkowski Frangetto, Ana Paula Beber Veiga and Gustavo Luedmann (eds.) *Legacy of the CDM: Lessons Learned and Impacts from the Clean Development Mechanism in Brazil as Insights for New Mechanisms* (IPEA 2019) 102.

7 UNFCCC 'Project Activities' <https://cdm.unfccc.int/Statistics/Public/CDMinsights/index.html>.

8 Non-Annex I countries are mostly developing countries. UNFCCC, *Achievements of the Clean Development Mechanism: Harnessing Invectives for Climate Action* (2018) <https://unfccc.int/sites/default/files/resource/UNFCCC_CDM_report_2018.pdf>.

9 UNFCCC (n 8) 2.

10 UNFCCC (n 8) 5.

More than 840,000 people were provided with access to clean drinking water. More than 150 million trees were planted.

Other notable positive impacts of CDM projects are related to the transfer of technologies to the developing world. While the exact impact of technology transfer is difficult to measure and quantify, certain studies have confirmed that, at least to a certain extent and under certain conditions, CDM projects have also resulted in technology transfer to the developing world.[11]

At the same time, there is notable criticism concerning the effectiveness and the implementation of the CDM. For example, many have questioned the actual extent of the emission reductions that were achieved because of CDM projects, as well as their 'additionality'.[12] Others raised doubts concerning the contribution of CDM projects to wider sustainable development goals, as well as regarding the utilisation of Official Development Assistance (ODA) funds for the promotion of CDM projects[13]: the prospect of public funds being re-directed from projects dealing with hunger and education in the poorest parts of the world, and given instead to rich energy corporations, is, admittedly, hard to digest. The impact that CDM projects' had on human rights is another oft-discussed topic. It is reported in this respect that certain CDM projects have resulted with the flooding of indigenous communities' lands and homes, forced relocation,[14] environmental destruction, as well as other human rights violations.[15]

All of these issues, and the fact that the KP itself did not address them in any meaningful way, cannot be ignored. These criticisms emphasise the need for a holistic, PCD-oriented approach, as will be addressed further

11 Kevin Murphy et al. 'Technology transfer in the CDM: An updated analysis' (2015) 15(1) *Climate Policy* 127; Gisele Schmid 'Technology transfer in the CDM: The role of host-country characteristics' (2012) 12(6) *Climate Policy* 722.

12 For example, it is debatable whether the Kyoto Protocol indeed led for reductions in GHG, see opposing views in Christian Almer and Ralph Winkler 'Analysing the effectiveness of international environmental policies: The case of the Kyoto Protocol' (2017) 82 *Journal of Environmental Economics and Management* 125.

13 Joyeeta Gupta, Harro van Asselt and Michiel van Drunen 'Global governance: Climate cooperation' in Joyeeta Gupta and Nicolien van der Grijp (eds.) *Mainstreaming climate change in development cooperation* (CUP 2010) 155–161.

14 Ariadni Chatziantoniou and Kelsey Alford-Jones 'Panama withdraws problematic Barro Blanco Project from CDM Registry' (December 2016) CIEL Blog <https://www.ciel.org/panama-withdraws-problematic-barro-blanco-dam-project-cdm-registry/>.

15 CIEL 'Lessons from CDM-registered Alto Maipo hydroelectric project: Impacted communities emphasise that any agreement on climate action must protect human rights & the environment' (December 2019) CIEL Press Room <https://www.ciel.org/news/lessons-from-cdm-registered-alto-maipo-hydroelectric-project-impacted-communities-emphasize-that-any-agreement-on-climate-action-must-protect-human-rights-the-environment/>; see more case-studies in Wolfgang Obergassel et al. 'Human rights and the Clean Development Mechanism: Lessons learned from three case studies' (2017) 8(1) *Journal of Human Rights and the Environment* 51.

below. At the same time, these problems did not detract from the fact that, *in principle*, the CDM model can be useful, even if important modifications will have to be made.

B The Paris Agreement

Surprisingly, the terms 'renewable energy' and 'fossil fuels' do not appear in the Paris Agreement. Just like in the case of the UNFCCC, also here the lack of explicit language on this issue should not be interpreted as a lack of attention. For example, Article 2 of the Paris Agreement is instructing the world's nations to ensure 'finance flows consistent with a pathway towards low greenhouse gas emissions and climate-resilient development'. This provision implies investment in, *inter alia,* mitigation measures, including the transition to renewable sources of energy.

Another relevant provision is Art 4, which sets up the Nationally Determined Contribution (NDC) mechanism. Under this Article, states' targets and plans shall be set, and updated, every five years. RE policies and targets are a major and inevitable element in states' NDCs, making the relevance of this Article to the promotion of RE policies (and RE-related investment more specifically) abundantly clear. A review of states' submitted NDCs indeed reveals that the vast majority of these plans include RE-related targets.[16] Moreover, the fact that most of these targets are conditional on external support[17] points at a crucial link with Foreign Direct Investment (FDI).

Perhaps the most promising and direct link between the Paris Agreement and RE-related investment in the developing world is found in Article 6. This Article instructs states to establish '[a] mechanism to contribute to the mitigation of greenhouse gas emissions and support sustainable development'. This mechanism – which since has been widely referred to as the 'Sustainable Development Mechanism' (SDM) – is supposed to achieve four goals:

(a) the promotion of GHG mitigation ('while fostering sustainable development')

(b) the incentivisation of private investment in activities that result in the mitigation of GHG

16 IRENA, *Renewable Energy in National Climate Action: Updates to IRENA's 2017 Analysis of the Renewable Energy Components of NDCs* (2018) <https://www.irena.org/publications/2018/Dec/Renewable-Energy-in-National-Climate-Action>.

17 See table with regional data on NDCs in IRENA 2018 (n 16) 4. According to this table, out of the 97 additional GW envisioned by Africa's NDCs, 55 GW are defined as 'conditional' defined as 'depend[ing] on international finance and other forms of assistance'. See IRENA, *Assessment of the Renewable Energy Components in Nationally Determined Contributions: Methodology* (IRENA 2018) <https://www.irena.org/-/media/Files/IRENA/Agency/Publication/2017/Nov/IRENA_NDC_methodology_2018.pdf?la=en&hash=4E6E9173BB306CDD4295F1E7F5FB1CF477CEAAF5> 11.

(c) contribute to the host state's emission reduction and (just like the KP's flexible mechanisms) allow other states to take credit for these reductions

(d) deliver global mitigation of GHG

At the time of writing, the specific details and rules concerning the shape and the operation of the SDM have not yet been agreed on. Nevertheless, the resemblance between this mechanism's skeleton and the KP's flexible mechanism is clear and intentional[18]: it is to be designed based on the '[e]xperience gained with and lessons learned from existing mechanisms and approaches adopted under the Convention and its related legal instruments'.[19] Notable features with special relevance for this study are the idea to incentivise the private sector's operation abroad, the mechanism's reliance on carbon markets,[20] as well as the possibility of crediting home states with GHG reductions achieved elsewhere. Just like in the case of the KP, these features are likely to result in the adoption of international law HCMs that will support RE-related FDI in the developing world.

III The UNFCCC and PCD

It is clear that the field of environmental law is not isolated from other subject-areas and that at least some appetite for promoting linkages with additional subject-areas exists.[21] Environmental law-based HCMs such as the CDM and the SDM are straightforward examples, reflecting an attempt to promote simultaneously environmental protection and developmental goals. Indeed, Art 12 of the KP (CDM) explicitly states that '[t]he purpose of the clean development mechanism shall be to assist Parties not included in Annex I in achieving sustainable development'.[22]

At the same time, it is evident that far too many linkages are still ignored and unexplored. As discussed below, environmental law HCMs such as the CDM do not rely on, or even refer to, international economic law. The KP's

18 The SDM's negotiations originated from a proposal to extend the CDM post-2020. See José Domingos Gonzalez Miguez and Tulio César Mourthé de Alvim Andrade 'The continuation of the CDM under the Paris Agreement and its articulation with the SDM' in Frangetto et al. (n 6) 293.

19 UNFCCC, Decision 1/CP.21, UN Doc. FCCC/CP/2015/10/Add.1, Para 37(f).

20 Art 6(2) and 6(4) of the Paris Agreement are referring to Internationally Transferred Mitigation Outcomes (ITMOs) as carbon units that will be sold on carbon markets.

21 The UNFCCC framework is indeed addressing a range of different climate-related elements, and increasingly so. For example, climate-induced migration was regarded for many years as falling outside of the UNFCCC's jurisdiction. Since 2011, however, this topic is mentioned in UNFCCC decisions and following the Paris Agreement the UNFCCC even created a multi-institutional Task Force in order to examine the link between climate change and migration.

22 Art 12, Kyoto Protocol.

lack of interaction with social elements (e.g. human rights, migration, poverty reduction) has also been a long-standing subject of criticism.[23]

The following section will examine some of these interactions in more detail. This examination will focus more closely on the CDM and the SDM mechanisms, which represent the main and most striking examples of environmental law HCMs in the context of RE-related investment.

A The UNFCCC, PCD and RE-related investment

The KP was the UNFCCC's first and most significant attempt to design an operational legal regime for the mitigation of climate change. There is little doubt that the KP has failed *politically*, primarily due to the lack of political will to undertake binding emission reductions.[24] This political failure, however, should not cast a shadow on the usefulness of the KP's economic and flexible mechanisms. These mechanisms, according to Bodansky et al., are 'likely to leave an enduring imprint on the climate regime'.[25]

The KP's mechanisms should not be discarded, and indeed, it seems that this will not be the case. Developing countries are interested in preserving the CDM or at least maintaining its main features. As discussed above, the legal basis for the CDM's likely successor – the SDM – is found in Art 6(4) of the Paris Agreement. The re-design of the CDM as a *Sustainable Development* Mechanism should be regarded as an opportunity to inject PCD into this legal tool and improve its sustainability and efficiency.

The starting point for any such reform is a difficult one. While CDM projects are indeed supposed to achieve broader 'sustainable development' outcomes,[26] the legal requirements that were attached to this instruction are very weak: the aspired 'sustainable development' standards were defined by (and often watered down by) host states with no international guidance,[27] and the enforcement of these standards was based on self-reporting, with very limited monitoring.[28]

23 Terms such as 'poverty' or 'human rights' do not appear in the KP's text. See more on this topic in this chapter below.

24 Authors such as William Nordhaus are attributing the failure of the KP also to elements such as its voluntary nature and the possibility to free ride the efforts of other states. William Nordhaus 'Climate clubs: Overcoming free-riding in international climate policy' (2015) 105(4) *American Economic Review* 1339.

25 Daniel Bodansky et al., *International Climate Change Law* (CUP 2017) 207.

26 See Article 12(2) Kyoto Protocol.

27 See further criticism in Karen Holm Olsen et al. 'Sustainability labelling as a tool for reporting the sustainable development impacts of climate actions relevant to Article 6 of the Paris Agreement' (2019) 19(2) *International Environmental Agreements: Politics, Law and Economics* 225.

28 See review in Axel Michaelowa, Aglaja Espelage, and Stephan Hoch 'Co-benefits under the market mechanisms of the Paris agreement' in Wolfgang Buchholz et al. (eds.) *Ancillary benefits of Climate Policy* (Springer 2020) 51.

The rules on the operation of the SDM have not yet been concluded, and it is not certain that the problems that plagued the CDM will feature also in the forthcoming SDM. Certain negative indications, however, already exist. Proposals to ensure that SDM projects will conform to the UN Sustainable Development Goals (SDGs) were flatly rejected.[29] Moreover, for the time being, terms such as 'sustainable development' and 'environmental integrity' remain undefined. The lack of explicit definitions could imply a watered-down implementation of these terms, as was the case with the term 'sustainable development' under the CDM. The fact that the Member States are not exploring potential linkages with economic legal instruments is another reason for concern.

On the other hand, not all of the indications are bleak. The Paris Agreement itself is peppered with references to 'sustainable development', and more specifically to social goals such as the protection of human rights, poverty reduction and more.[30] These references indicate at least a certain attentiveness to the criticism expressed in this chapter concerning the CDM's social and environmental externalities, and it is very likely that at least some of this language will find its way into the core of the SDM rules, in one form or another.[31] At the same time, it is difficult to ignore the fact that at the time of writing – some years after the conclusion of the Paris Agreement – available details on how such reforms will materialise have not been introduced. These delays are telling on their own.

Below is a more detailed evaluation of the interactions between the CDM, the SDM, and other related international law HCMs.

i The climate regime and economic law

In its current form, there is no doubt that the CDM is lacking in PCD. The most striking example is the lack of interaction with other economic law instruments. This is especially problematic as the overlap in the regulated

29 Sandra Greiner et al., *Moving Towards Next Generation Carbon Markets: Observations from Article 6 Pilots* (Climate Focus 2019) 42.

30 The preamble to the Paris Agreement states that the Parties are

> *acknowledging* that climate change is a common concern of humankind, Parties should, when taking action to address climate change, respect, promote and consider their respective obligations on human rights, the right to health, the rights of indigenous peoples, local communities, migrants, children, persons with disabilities and people in vulnerable situations and the right to development, as well as gender equality, empowerment of women and intergenerational equity.

Articles 2, 4 and 6 are addressing the eradication of poverty, and the Agreement's text includes 12 references to the term 'sustainable development'.

31 For example, Art 6(2) is stating in this respect that the future SDM will 'promote sustainable development and ensure environmental integrity and transparency, including in governance […]. See Art 6(2) (and also 6(1)) of the Paris Agreement.

subject-matter is vast and remarkable. Investment treaties, for example, could play a supportive role in CDM projects by providing investors with a stable and predictable legal environment, protection from expropriation and depoliticised forms of dispute settlement. Werksman et al. noted already in 2003 that 'investment law, appropriately constructed, can also support the effective functioning of the CDM'.[32] Authors such as Baetens and Boute have mentioned the possibility of incorporating investment protection instructions into the KP.[33] Edna Sussman dismissed the possibility that such a new regime be negotiated under the UNFCCC framework as 'impractical', but proposed instead that the Energy Charter Treaty (ECT) – including this treaty's investment chapter – should be linked to the KP.[34] The ECT, she explains, already refers to the UNFCCC in its preamble, and thus (per the Art 31 of the Vienna Convention on the Law of Treaties), this treaty's investment protection instructions should be interpreted and understood in light of the UNFCCC's objectives and instructions.[35]

Other missing linkages concern trade laws and policies. For example, Schmid has evaluated the impact of CDM projects on the process of technology transfer.[36] This author identified restrictive trade policies (notably tariffs on Environmental Goods and Services (EGS)) as an inhibiting factor that adversely affects the process of technology transfer through CDM projects.[37] Other authors have mentioned barriers such as the lack of 'trained national CDM consultants'[38]: a problem that could be addressed through the entrance of actors such as international consultancies via the liberalisation of trade in services.

Others have warned of an additional problem that certain trade rules might conflict with the CDM mechanism, including rules on subsidies, technical barriers and agriculture,[39] as well as the WTO's rules in investment

32 Jacob Werksman, Kevin Baumart and Navroz Dubash 'Will international investment rules obstruct climate protection policies? (2003) 3 *International Environmental Agreements: Politics, Law and Economics* 59.

33 Freya Baetens 'The Kyoto protocol in investor-state arbitration: Reconciling climate change and investment protection objectives' in Marie-Claire Cordonier Segger, Markus Gehring and Andrew Newcombe (eds.) *Sustainable Development in World Investment Law* (Kluwer 2011) 707; Anatole Boute 'Combating climate change through investment arbitration' (2012) 35(3) *Fordham International Law Journal* 614.

34 Edna Sussman 'The energy charter treaty's investor protection provisions: Potential to foster solutions to global warming and promote sustainable development' in Cordonier Segger et al. (n 138) 529.

35 Sussman (n 34).

36 Schmid (n 11).

37 Schmid (n 11) part 4.2.

38 Paula Castro and Axel Michaelowa 'Would preferential access measures be sufficient to overcome current barriers to CDM projects in least developed countries? (2011) 3(2) *Climate and Development* 123, 126.

39 Thomas Brewer 'The WTO and the Kyoto Protocol: Interaction issues' (2004) 4(1) *Climate Policy* 3.

(TRIMS).[40] Most will agree, however, that a careful design of CDM-related regulations can prevent these conflicts,[41] and past experience indeed demonstrates that CDM trade-related disputes have not materialised to date: a conclusion that probably downplays any claims concerning potential conflicts. This discussion of conflicts between WTO Law and the KP is beyond the scope of this chapter, and this study more generally. The drafters of the prospective SDMs (and related domestic regulations) will nevertheless do well to address these interactions and avoid potential pitfalls.[42]

ii The climate regime and environmental policies and objectives

The KP's flexible mechanisms have been criticised in the past for their negative impact on certain environmental issues. There are different types of externalities that require attention in this respect. First, the most common criticism of the CDM/JI addresses these mechanisms' actual contribution to global emission reductions. It is often questioned whether the reduction in emissions brought about by these projects would not have happened regardless of the KP, or whether the calculations of benefits have been overestimated. Some even went as far as arguing that '[i]t is likely that the large majority of the projects registered and CERs issued under the CDM are not providing real, measurable and additional emission reductions'.[43] These problems, however, concentrate on the inefficiencies within the climate regime itself, rather than pointing at any lack of PCD.

Of more interest are the linkages between the CDM (a climate/development tool) and other non-climate environmental goals. Arens and Mersmann have identified a list of areas that can potentially be affected by these projects, and advocated the establishment of appropriate safeguards under Art 6 of the Paris Agreement.[44] These include mostly social issues (discussed in the part below) *inter alia* human rights, labour rights, access to services, the rights of marginal/vulnerable groups, women's rights, indigenous people

40 Stefan Rechsteiner, Christa Pfister and Fabian Martens 'TRIMS and the clean development mechanism – potential conflicts' in Thomas Cottier, Olga Nartova and Sadeq Bigdeli (eds.) *International Trade Regulation and the Mitigation of Climate Change* (CUP 2009) 299.

41 Annie Petsonk 'The Kyoto protocol and the WTO: Integrating greenhouse gas emissions allowance trading into the global marketplace' (1999) 10 *Duke Environmental Law & Policy Forum* 185.

42 Notably discriminatory policies and the use of local content requirements should be addressed with caution.

43 Martin Cames et al., *How Additional Is the Clean Development Mechanism? Analysis of the Application of Current Tools and Proposed Alternatives* (Öko-Institut 2016) 11.

44 Christof Arens and Florian Mersmann 'Positive results, no negative consequences: No harm options for Article 6' (2018) JIKO Policy Paper No. 03/2018 <https://epub.wupperinst.org/frontdoor/deliver/index/docId/7123/file/7123_Article6.pdf> 2–3.

rights, forced settlement, public health and physical and cultural heritage.[45] With respect to environmental factors, these authors identified as relevant the conservation of bio-diversity, natural habitats, air pollution, waste, land and soil conservation.[46]

One oft-discussed example regards windfarms and the multitude of impacts that these have on wildlife. The construction of offshore windfarms, for example, can adversely impact marine mammals and fish (impact from noise was stressed in this process),[47] while their operation (as the operation of onshore windfarms[48]) can lead to birds' mortality and displacement.[49] The impact of solar farms on wildlife is not as well founded as that of windfarms.[50] Concerns regarding the extensive land-use that increased reliance on solar farms requires, however, were raised by some with respect to eco-systems[51] and biodiversity.[52] The UK's Royal Society for the Protection of Birds (RSPB) has stressed the importance of location in terms of risks for birds.[53] The RSPB warned of risks posed by solar farms for birds (including collusion and burns), insects,[54] as well as with respect to wildlife displacement and loss of habitats.[55]

Whether the term 'environmental integrity', as aspired by the Art 6 of the Paris Agreement, is indeed addressing these non-climate environmental problems is unclear as the exact rules have not yet been agreed. A possible interpretation of this term could be drawn from the Paris Agreement's preamble,[56] according to which the Parties are:

45 Arens and Mersmann (n 44) 2–3.
46 Arens and Mersmann (n 44) 2–3.
47 Lena Bergström et al. 'Effects of offshore wind farms on marine wildlife—A generalized impact assessment' (2014) 9(3) *Environmental Research Letters* 5.
48 Chris Thaxter 'Birds and bat species' global vulnerability to collusion mortality at wind farms revealed through a trait-based assessment' (2017) 13 *Proceedings of the Royal Society* B 284.
49 Robert Furness, Helen Wade and Elisabeth Masden 'Assessing vulnerability of marine birds populations to offshore farms' (2013) 119 *Journal of Environmental Management* 56.
50 Chris Harrison, Huw Lloyd and Chris Field, *Evidence Review of the Impact of Solar Farms on Birds, Bats and General Ecology* (2017 Natural England) 13.
51 Antoine Beylot et al. 'Environmental impacts of large-scale grid connected ground-mounted PV installations' (2014) 61 *Renewable Energy* 2, 5.
52 A Natural England report mentions potential impact on wildlife (if solar farms are located in sites 'of high wildlife value') but admits that '[v]ery little research has been conducted to date'. See Natural England 'Solar parks: maximising environmental benefits' (2011) Technical Information Note TIN101 <https://webarchive.nationalarchives.gov.uk/20150902172007/http://publications.naturalengland.org.uk/publication/32027> 3.
53 Royal Society for the Protection of Birds (RSPB) 'Solar energy' (December 2014) RSPB Policy Briefing <http://ww2.rspb.org.uk/Images/Solar_power_briefing_tcm9-273329.pdf> 3–4.
54 RSPB (n 53) 3–4.
55 RSPB (n 53) 4–5.
56 According to Art 31 of the 1969 Vienna Convention on the Law of Treaties, a treaty should be interpreted in light of *inter alia* its preamble.

> Noting the importance of ensuring the integrity of all ecosystems, including oceans, and the protection of biodiversity, recognized by some cultures as Mother Earth, and noting the importance for some of the concept of "climate justice", when taking action to address climate change.

According to some, this holistic view is indeed supported by some states.[57] Despite this possibility, it seems that the position of most negotiating states is that the 'environmental integrity' requirement will address only climate-related considerations,[58] a point that contradicts the notion of PCD.

iii The climate regime and the social legal sphere

Perhaps the most discussed PCD-related interaction is between climate policies and social development. Indeed, attempts to integrate social development within the UNFCCC framework are increasing. As a point of departure, the UNFCCC itself is relying on 'sustainable development' as one of its leading principles[59]: a holistic term that embeds the imperative to integrate social well-being into the process of development. This Convention further states in this respect that the Member States are '[a]ffirming that responses to climate change should be coordinated with social and economic development in an integrated manner with a view to avoiding adverse impacts on the latter'.[60] The KP similarly asks the Parties to assess 'the overall effects of the measures taken pursuant to this Protocol, in particular environmental, economic and social effects'.[61] The Paris Agreement's preamble likewise reminds the Parties that they 'may be affected not only by climate change, but also by the impacts of the measures taken in response to it'.[62]

While these references are certainly important, one must admit that they are also vague and lacking in concrete application. As discussed above, the implementation of these references in the context of CDM was heavily criticised. The link between environmental law HCMs and human rights is especially important in this respect and worth recalling. This link is also interesting as a case study, addressing the manner in which social development should be incorporated into climate change policies, and how, in

57 Carbon Market Watch 'Building blocks for a robust sustainable development mechanism' (2017) Carbon Market Watch Policy Brief <https://carbonmarketwatch.org/wp-content/uploads/2017/05/BUILDING-BLOCKS-FOR-A-ROBUST-SUSTAINABLE-DEVELOPMENT-MECHANISM_WEB-SINGLE_FINAL.pdf> 5.

58 Asian Development Bank, *Decoding Article 6 of the Paris Agreement* (ADB 2018) 9–10.

59 UNFCCC, Art 3(4).

60 UNFCCC, Preamble.

61 Kyoto Protocol, Art 13(4)(a).

62 Paris Agreement, Preamble.

practice, the linking of these subject-areas was carried out. The following part will address this interaction in more detail.

iv Interaction between climate policies and the protection of human rights

There are numerous reported examples of cases in which CDM projects resulted in human rights violations.[63] Some of these projects ended up inflicting human rights violations, including the expropriation of lands from local communities, forced evictions and consequent violent response towards these communities when they tried to resist.[64] Indeed, both former and current UN Special Rapporteurs on Human Rights and the Environment – Professors John Knox and David Boyd – have stressed the importance of integrating human rights considerations into the design of Art 6's future mechanisms.[65]

v Environmental law HCMs and human rights: in search for PCD

While the link between environmental law HCMs and human rights is well established, it seems that the legal mechanisms are still fragmented. For almost 20 years, the UNFCCC framework mostly ignored the link between climate change policies and human rights. Indeed, neither the UNFCCC nor the KP explicitly mentions the term 'human rights'.[66] As reported by some,[67] a change in approach was finally triggered by the Office of the High

63 See review in Jeanette Schade and Wolfgang Obergassel 'Human rights and the clean development mechanism' (2014) 27(4) *Cambridge Review of International Affairs* 717, 724; Damilola Olawuyi 'Fostering accountability in large-scale environmental projects: Lessons from CDM and REDD+ Projects' (2015) 6 *The World Bank Legal Review: Improving Delivery in Development: The Role of Voice, Social Contract, and Accountability* 129.

64 Schade and Obergassel (n 63) ibid.

65 Knox (n 77); UNGA, *Report of the Special Rapporteur on the Issue of Human Rights Obligations Relating to the Enjoyment of a Safe, Clean, Healthy and Sustainable Environment* (2019) un Doc A/74/161 <https://undocs.org/en/A/74/161> para 69.

66 The wider 'sustainable development' imperative is mentioned, including an obligation to 'minimize adverse effects, including the adverse effects of climate change, effects on international trade, and social, environmental and economic impacts [...]. See Art 2(3) of the Kyoto Protocol.

67 Bodansky et al. (n 25) 297. Knox attributes the change in approach to earlier events, notably the 2005 Petition to the Inter-American Commission on Human Rights Seeking Relief from Violations Resulting from Global Warming Caused by Acts and Omissions of the United States, in which Inuit communities claimed that the US GHG pollution is leading to human rights violations. See John Knox 'The Paris agreement as a human rights treaty' in Dapo Akande et al. (eds.) *Human Rights and the 21st Century Challenges: Poverty, Conflicts and the Environment* (OUP Forthcoming) <https://papers.ssrn.com/sol3/papers.cfm?abstract_id=3192106&download=yes>.

Commissioner of Human Rights and its influential 2009 report[68] on the link between human rights and climate change. This report has led to several developments within the UN system,[69] one of which is the text of the 2010 Cancun Agreements where the Parties recognised the impact that climate change may have on human rights. The Paris Agreement was the next important step,[70] stating that the Parties are:

> Acknowledging that climate change is a common concern of humankind, Parties should, when taking action to address climate change, respect, promote and consider their respective obligations on human rights, the right to health, the rights of indigenous peoples, local communities, migrants, children, persons with disabilities and people in vulnerable situations and the right to development, as well as gender equality, empowerment of women and intergenerational equity.

Interestingly, the term 'human rights' is mentioned in the Paris Agreement only once (in this treaty's preamble, quoted above). This fairly limited reference was enough for the former UN Rapporteur on Human Rights and the Environment, John Knox, to determine that 'the Paris Agreement is also a human rights treaty'.[71] Knox further explains[72]:

> Of course, the Paris Agreement is not a human rights treaty in the usual sense. It does not define rights or create mechanisms to review and promote their compliance. Nevertheless, it makes an important contribution to the fulfilment of human rights. By clarifying that actions to address climate change should take human rights into account, the agreement helps to mainstream human rights norms into the ongoing implementation and evolution of the climate regime.

While the reference to the term 'human rights' is indeed important, one may question Knox's defining of this Agreement as a 'human right treaty'. Indeed, in light of Knox's explanation, it seems that he himself will admit that this classification was somewhat of a rhetorical exaggeration. The reference to human rights in the Paris Agreement is declaratory in nature, and it is doubtful whether it changes anything in states' legal obligations. Just as it would be wrong to argue that human rights laws were not applicable to

68 OHCHR, *Report of the Office of the United Nations High Commissioner for Human Rights on the relationship between Climate Change and Human Rights* (2009) A/HRC/10/61.

69 Notably UN Human Rights Council Resolution 10/4 on Human Rights and Climate Change.

70 A full review of the increased interaction between human rights and climate change is beyond the scope of this study and is available in Knox (n 67).

71 Knox (n 67).

72 Knox (n 71).

states' climate policies before the Paris Agreement, it is equally erroneous to argue that additional norms will now have to be applied. Also the fact that the term 'human rights' is mentioned only once in the Paris Agreement (in its *non-binding* preamble) and the fact that a much wider recognition of this term was explicitly rejected in earlier drafts[73] are both very telling, and provide the reader with more perspective.

A number of organisations have continued to stress the link between climate policies and human rights. An important example is a joint-statement that was issued by five UN bodies on this topic.[74] This statement as well can be seen as questioning Knox's view that the Paris Agreement indeed changed the role and place of human rights, as it is simply 'urging' 'all States to take into consideration their human rights obligations as they review their climate commitments'.[75] In other words, these organisations recognise that nothing has really changed, and that states needed to be reminded of their existing obligations. This statement nevertheless follows the crux of Knox's argument (which remains uncontested in this study), and stresses the importance of the integration of human rights within climate policies:

> When reducing emissions and adapting to climate impacts, States must seek to address all forms of discrimination and inequality, including advancing substantive gender equality, protecting the rights of indigenous peoples and of persons with disabilities, and taking into consideration the best interests of the child. [...] should offer complementary protection mechanisms and temporary protection or stay arrangements for migrant workers displaced across international borders in the context of climate change or disasters and who cannot return to their countries [...] respect, protect and fulfil the rights of all, including by mandating human rights due diligence and ensuring access to education, awareness raising, environmental information and public participation in decision-making. In particular, States have the responsibility to protect and defend effectively the rights of environmental human rights defenders, including women, indigenous and child environmental defenders.[76]

73 Julia Dehm 'Reflections on Paris: Thoughts towards a critical approach to climate law' (2018) 1(1) *Revue Québécoise de droit international* 61, 88.

74 The relevant bodies are the Committee on the Elimination of Discrimination Against Women, the Committee on Economic, Social and Cultural Rights, the Committee on the Protection of the Rights of All Migrant Workers and Members of their Families, the Committee on the Rights of the Child and the Committee on the Rights of Persons with Disabilities. See Joint Statement on "Human Rights and Climate Change" 16 September 2019, <https://www.ohchr.org/en/NewsEvents/Pages/DisplayNews.aspx?NewsID=24998&LangID=E>.

75 Joint Statement on Human Rights and Climate Change (n 74) para 1.

76 Joint Statement on Human Rights and Climate Change (n 74) paras 4–7.

These developments are encouraging as they stress the importance of improved PCD and the necessity to design a more comprehensive legal framework. These declarations, however, also demonstrate the problematic legal reality: despite improved awareness, the legal reality remains unchanged. As the SDM is intended to be more sustainable and holistic than its predecessor, it is hoped that its design and details will reflect a better linkage with human rights obligations and provide better safeguards.

vi Integrating human rights into the SDM: Knox's blueprint

The disappointment from the unsatisfactory integration of human rights safeguards into the CDM legal framework implies that the design of the SDM will have to be improved. Knox provides useful guidelines as for what the integration of human rights law into the future SDM should include. Knox's guidelines are crucial for the purpose of this study: they present a blueprint for the necessary safeguards that any RE-related international law HCM will have to include. These guidelines are provided in the form of a 'checklist'[77]:

1 Providing for environmental assessment and making environmental information public;
2 Facilitating public participation in environmental decision-making, including by protecting the rights of freedom of expression and association;
3 Providing for effective remedies for violations of human rights;
4 Adopting and implementing effective legal and institutional frameworks to address environmental harm;
5 Providing heightened protections for those who are particularly vulnerable; and
6 Respecting and protecting all other relevant human rights.

The following section will review the 'items' on this checklist one by one. As will be demonstrated below, the use of these safeguards is not uncommon in international law and there are numerous models from which treaty negotiators will be able to draw inspiration and important background support. The abundance of examples shows also that the integration of most of these safeguards is perhaps more common and far less controversial than one may have assumed: the next step (i.e. adding these to the SDM) need therefore not necessarily meet with strong resistance. Other features, however, will require more imagination and, not less importantly, an increased political will.

77 John Knox 'Human rights and safeguards in the new climate mechanism established in Article 6, paragraph 4 of the Paris agreement' (2016) <https://www.ohchr.org/Documents/Issues/Environment/Letter_to_SBSTA_UNFCCC_May2016.pdf> 3–4.

ITEM 1: ENVIRONMENTAL AND SOCIAL IMPACT ASSESSMENT STUDIES

Environmental Impact Assessments (EIAs) are future-facing studies that are aimed at predicting the projects' future impact on the environment. EIAs are required today by a large number of international treaties,[78] and, in certain jurisdictions, also before the approval of new investment projects (e.g. EU,[79] Canada[80]). Knox posits the use of EIA as a fundamental preventive safeguard, which states must rely on in order to ensure the effective protection of human rights.[81]

Traditionally, EIAs were relied on for the purpose of predicting potential impact on the environment. In recent years however, the coverage of EIAs is increasing and social elements – notably impact on human rights – are examined as well. While according to some the inclusion of human rights in EIAs is not yet widespread,[82] there is no doubt that a significant pool of useful examples nevertheless exist, on both the national and international levels. For example, Canadian law requires the examination of projects' sustainability (understood as 'the ability to protect the environment, contribute to the social and economic well-being of the people of Canada'[83]), including several specific references to indigenous people's rights.[84] The EU's practice on the international level (i.e. before the conclusion of trade and investment treaties) is more explicit, demanding an examination of human rights,[85] in accordance with the EU Treaties' provisions.[86]

78 See for example the Espoo Convention on Environmental Impact Assessment in a Transboundary Context (1997); Art 14 of the Convention on Biological Diversity (1992); Art 206 of the United Nations Convention on the Law of the Sea, and Principle 17 of the Rio Declaration on Environment and Development; Art 24(7) of the USMCA. See full review in David Collins 'Public participation in environmental impact assessments for foreign investment projects: A Canadian perspective' in de Brabandere, Gazzini and Kent (eds.) (n 106).

79 Directive 2014/52/EU of the European Parliament and of the Council of 16 April 2014 amending Directive 2011/92/EU on the assessment of the effects of certain public and private projects on the environment.

80 Canada Impact Assessment Act (2019).

81 Knox (n 77) 6.

82 Nora Götzmann 'Human rights impact assessment of business activities: Key criteria for establishing a meaningful practice' (2017) 2(1) *Business and Human Rights Journal* 87.

83 *Practitioner's Guide to Federal Impact Assessments under the Impact Assessment Act* <https://www.canada.ca/en/impact-assessment-agency/services/policy-guidance/practitioners-guide-impact-assessment-act.html>, see under 'Considering the extent to which a project contributes to sustainability' <https://www.canada.ca/en/impact-assessment-agency/services/policy-guidance/practitioners-guide-impact-assessment-act/interim-guidance-considering.html>.

84 Ibid.

85 See for example the EU's practice, in European Union, *Handbook for Trade: Sustainability Impact Assessment* (EU 2016) <https://trade.ec.europa.eu/doclib/docs/2016/april/tradoc_154464.PDF>.

86 Notably Art 21 of the Treaty on European Union, according to which

> The Union's action on the international scene shall be guided by the principles which have inspired its own creation, development and enlargement, and which it seeks to

Another example is the Activity of the United States' Overseas Private Investment Corporation (OPIC), recently merged into the nascent US International Development Finance Corporation (DFC). According to the DFC's Environmental and Social Policy and Procedures,[87] risky international projects[88] wishing to receive support from DFC/OPIC are subjected to a full Environmental and Social Impact Assessment (ESIA). OPIC's Environmental and Social Policy Statement specifically confirmed that '[t]he term "social", as used throughout this Policy Statement, is inclusive of the labor and human rights dimensions of sustainable development'.[89] While the DFC's new guidance documents on ESIAs do not provide such a straightforward reference, it is nevertheless clear that human rights are to be evaluated in these assessments also following the assimilation of OPIC into DFC.[90] The introduction to DFC's Environmental and Social Policy and Procedures is explicit on this matter, stating that 'DFC investments adhere to high standards and respect the environment, human rights, and worker rights'.[91]

On the international level, the (non-binding) United Nations Guiding Principles (UNGP) on Business and Human Rights instruct that '[i]n order to prevent and mitigate adverse human rights impacts, business enterprises should integrate the findings from their impact assessments across relevant internal functions and processes, and take appropriate action'.[92] Another example from this area is the operation of the World Bank's Multilateral Investment Guarantee Agency (MIGA). MIGA's Performance Standards on Environmental and Social Sustainability require that[93]:

---

advance in the wider world: democracy, the rule of law, the universality and indivisibility of human rights and fundamental freedoms, respect for human dignity, the principles of equality and solidarity, and respect for the principles of the United Nations Charter and international law.

87 See para 3.10.

88 See para 3.10 of the DFC Environmental and Social Policy and Procedures https://www.dfc.gov/sites/default/files/media/documents/DFC_ESPP_012020.pdf 'Applicants seeking support for category A and B projects, and special consideration projects, must conduct some form of environmental and social assessment.'; Guidance Note 1 Social and Environmental Assessment and Management Systems <https://www.ifc.org/wps/wcm/connect/ab6cffa1-e357-42e3-a27b-c759ffa561a1/2007%2BUpdated%2BGuidance%2BNote_1.pdf?MOD=AJPERES&CVID=jqeu0U7> G36.

89 OPIC, Environmental and social policy statement <https://www.dfc.gov/sites/default/files/2019-08/consolidated_esps.pdf>.

90 See Annex A of Guidance Note 1 (n 88) for a review of the requirements/content of such ESIA.

91 DFC Environmental and social policy and procedures (n 88).

92 United Nations guiding principles on business and human rights (2011).

93 Para 5, MIGA, Performance standards on environmental and social sustainability (2013) <https://www.miga.org/sites/default/files/archive/Documents/MIGA_Performance_Standards_October_2013.pdf>. Footnote 11 of this document states that

> [f]or greenfield developments or large expansions with specifically identified physical elements, aspects, and facilities that are likely to generate potential significant environmental or social impacts, the client will conduct a comprehensive Environmental

> The client, in coordination with other responsible government agencies and third parties as appropriate, will conduct a process of environmental and social assessment, and establish and maintain an ESMS appropriate to the nature and scale of the project and commensurate with the level of its environmental and social risks and impacts.

A further example is found in the Adaptation Fund's Environmental and Social Policy.[94] The Fund requires its implementing entities (notably development banks) to 'prepare an environmental and social assessment that identifies any environmental or social risks, including any potential risks associated with the Fund's environmental and social principles set forth above'.[95] The elements that require examination include a variety of environmental impacts (e.g. climate change, pollution, biodiversity and more[96]), and also a comprehensive list of human rights-related impacts, including on marginalised and vulnerable groups; involuntary resettlement; indigenous peoples; gender equity; and a broad, catch-all 'human rights' category.[97] More examples are found in other climate funds, including in the environmental and social policies of the Global Environmental Facility (GEF) and the Green Climate Fund (GCF), both of which are more recent (adopted in 2018).[98]

Lastly, the importance of EIAs was recognised also in the more direct context of development assistance. The Paris Declaration on Aid Effectiveness is calling for an increased reliance on EIAs, and commits, *inter alia*, to

> [s]trengthen the application of EIAs and deepen common procedures for projects, including consultations with stake holders; and develop and apply common approaches for "strategic environmental assessment" at the sector and national levels.[99]

On the face of it, given that the use of EIA (including the examination of human rights) is fairly acceptable in international law, the inclusion of

---

and Social Impact Assessment, including an examination of alternatives, where appropriate.

94 Adaptation Fund, *Environmental and Social Policy* (Approved in November 2013) <https://www.adaptation-fund.org/wp-content/uploads/2015/09/Environmental-Social-Policy-approved-Nov2013.pdf>.

95 Adaptation Fund, *Environmental and Social Policy* (n 94) para 30.

96 Adaptation Fund, *Environmental and Social Policy* (n 94) Part B.

97 Adaptation Fund, *Environmental and Social Policy* (n 94) Part B.

98 See in part 6.3 of the Green Climate Fund, *Environment and Social Policy* (GCF 2018) <<https://www.greenclimate.fund/sites/default/files/document/environment-social-policy.pdf>; and Minimum Standard 1 of the GEF, *Policy on Environmental and Social Safeguards* (2018) SD/PL/03 <<https://www.thegef.org/sites/default/files/documents/gef_environmental_social_safeguards_policy.pdf>.

99 *The Paris Declaration on Aid Effectiveness* (2005) Para 41.

instructions on the conduct of such studies under the future SDM should not be regarded as controversial. Such instructions should include an examination of potential environmental and social (including human rights) impacts by SDM projects, in line with one of the many models that are currently available. Examples for such models include those presented above or other models that were developed by non-profit institutions such as the Business for Social Responsibility (BSR)[100] or the Danish Institute for Human Rights,[101] both of which have compiled a list of guidelines and practical steps that Human Rights Impact Assessment (HRIA) could follow.[102] Proposals in this spirit were indeed made by scholars with respect to other features of the climate regime – *inter alia* the CDM and REDD+ mechanisms[103] – and more recently also specifically with respect to the SDM itself.[104]

It is true that a requirement to conduct EIAs will imply additional costs: these studies are usually commissioned to specialised consultancies, at a price. Nevertheless, given that these costs are likely to fall on the investor, and given that investors are already conducting a line of assessments before making their investment, it is unlikely that these costs will 'chill' investors and adversely affect the flow of investments. If anything, they will warn investors of future pitfalls, ensure the durability of their projects and prevent future disputes.

Perhaps a more significant hurdle for the mandatory inclusion of EIAs in the future SDM will be the political objections that are almost routinely made in light of any attempt to improve the protection of human rights. This obstacle may be especially problematic where HRIA are expected to evaluate also existing domestic legal frameworks and their effectiveness in protecting human rights (for example, when questioning whether sufficient safeguards and monitoring exist). It could be that states will not be so eager

100 BSR, *Conducting an Effective Human Rights Impact Assessment: Guidelines, Steps and Examples* (BSR 2013) <http://www.bsr.org/reports/BSR_Human_Rights_Impact_Assessments.pdf>.

101 The Danish Institute for Human Rights, *Introduction to Human Rights Impact Assessments: Guidance on the Fundamental Principles of HRIA and Why Businesses Need to Assess Their Human Rights Impacts* (The Danish Institute for Human Rights 2020) <https://www.humanrights.dk/tools/human-rights-impact-assessment-guidance-toolbox/introduction-human-rights-impact-assessment>.

102 A review of other models is available in World Bank, *Human Rights Impact Assessments: A Review of the Literature, Differences with Other Forms of Assessments and Relevance for Development* (Commissioned by the Nordic Trust Fund 2013) <http://siteresources.worldbank.org/PROJECTS/Resources/40940-1331068268558/HRIA_Web.pdf>.

103 See for example Damilola S. Olawuyi, *The Human Rights Based Approach to Carbon Finance* (CUP 2016) 345; Malayna Raftopoulos 'REDD+ and human rights: addressing the urgent need for a full community-based human rights impact assessment' (2016) 20(4) *The International Journal of Human Rights* 509.

104 Paola Villavicencio Calzadilla 'Human rights and the new sustainable mechanism of the Paris agreement: A new opportunity to promote climate justice' (2018) 21 *Potchefstroom Electronic Law Journal* <http://www.saflii.org/za/journals/PER/2018/56.pdf>.

to link their ability to attract lucrative energy-related investment with an ongoing review of their human rights policies and track record in this area.

On the other hand, as stated above, the fact that somewhat similar reviews are already being conducted by Intergovernmental Organizations (IGOs), including development banks, MIGA and others, may alleviate these concerns. Indeed, reports made by bodies such as MIGA have not so far attracted any significant public attention. States' fear from scrutiny in this respect may be exaggerated.

ITEM 2 (AND 1): PUBLIC PARTICIPATION AND ACCESS TO INFORMATION

Items 2 on Knox's list (combined with Item 1) mostly relates to the access to information and public participation. These elements have also been stressed by Knox's successor – Professor David Boyd – in his more recent report.[105] These legal features/rights are already implemented in many regions of the world and could be potentially linked to the SDM. There are several models from which states can benefit from previous experience in this respect, covering all stages of the investment life-cycle.[106]

The Aarhus Convention model The example of the Aarhus Convention and its enforcement mechanisms could be used as a model, for designing a new designated legal mechanism. This could be done through either the expansion of the Aarhus Convention itself, or the conclusion of a side-agreement that will be based on roughly similar instructions and mechanisms.

This Convention instructs *inter alia* the establishment of effective mechanisms that will facilitate the public's participation in decision-making processes and allow access to relevant information.[107] This Convention applies also to a list of energy-related activities: admittedly mostly nuclear and fossil fuels, but in some cases also to renewable projects.[108] Indeed, renewable energy projects were contested under the Aarhus mechanisms, where members of the public challenged issues such as insufficient

105 *Report of the Special Rapporteur* (n 65) Para 64.

106 For an elaborate study on this topic, see Eric de Brabandere, Tarcisio Gazzini and Avidan Kent (eds.) *Public Participation and Foreign Investment Law: From the Creation of Rights and Obligations to the Settlement of disputes* (Brill 2021).

107 Art 1, Aarhus Convention on access to information, public participation in decision-making and access to justice in environmental matters (1998).

108 Art 6(1)(a) instructs that the Parties 'shall apply the provisions of this article with respect to decisions on whether to permit proposed activities listed in annex I'. Annex I addresses the 'Energy Sector', and while it is not explicitly stating renewable energy projects in its list of covered activities, it does add that 'Any activity not covered by paragraphs 1–19 above where public participation is provided for under an environmental impact assessment procedure in accordance with national legislation'. This instruction *de facto* adds renewable energy projects in many parts of the world.

dissemination of information to the public,[109] or failure to allow effective public participation.[110]

The Aarhus model is becoming increasingly popular. There are currently 47 signatory states to this Convention. An additional 22 states have recently concluded a twin Latin American Convention, committing themselves to largely similar obligations.[111] Bearing in mind that somewhat similar (even if not identical) instruments exist also in other states' domestic and regional legal frameworks (e.g. the USMCA's Environment Chapter[112]), it is likely that wide agreement (even if not a full international consensus) on this matter will not be hard to achieve. Alternatively, states could also begin with a Plurilateral Side Agreement to the SDM (or an Annex) that will allow more flexibility for states.

Admittedly, following the Aarhus model will impose additional costs and bureaucracies: notably the creation of a secretariat and the operation of a complaint committee. While these costs could certainly be reduced through the use of existing institutions (e.g. the Aarhus secretariat, or existing UN-FCCC divisions), they should nevertheless be acknowledged if this model is to be pursued.

Public participation through impact assessment mechanisms Another mechanism through which public participation can be secured – and linked to the future SDM – is EIA studies. The role of EIAs (and HRIAs) has been discussed above in more detail. It will therefore be discussed in this part only briefly, and only in the context of public participation.

Most EIAs are relying heavily on public participation, an element considered by many as an indispensable condition for ensuring the effectiveness of the process and its legitimacy.[113] Glucker et al. explain in this respect that '[n]ot only is public participation in EIA a goal in itself, there seems to be widespread consensus that public participation is also key to effective environmental assessment'.[114]

109 Economic Commission for Europe 'Findings and recommendations with regard to communication ACCC/C/2010/54 concerning compliance by the European Union' (2012).

110 Economic Commission for Europe 'Findings and recommendations with regard to communication ACCC/C/2012/68 concerning compliance by the European Union and the United Kingdom of Great Britain and Northern Ireland' (2013).

111 Regional agreement on access to information, public participation and justice in environmental matters in Latin America and the Caribbean (concluded on 27 September 2018).

112 See several references in this chapter, interestingly Article 24.27 (the 'Submissions of Enforcement Matters' mechanism).

113 See for example the preamble to Canada's Impact Assessment Act (2019); para 16 in the preamble to the EU Environmental Impact Assessment Directive; Preamble to Chapter 23, Agenda 21; See review in David Collins 'Public participation in environmental impact assessments for foreign investment projects: A Canadian perspective' in de Brabandere, Gazzini and Kent (eds.) (n 106).

114 Anne Glucker, Peter Driessen, Arend Kolhoff and Hens Runhaar 'Public participation in environmental impact assessment: why, who and how? (2013) 43 *Environmental Impact Assessment Review* 104.

The importance of public participation in the preparation of impact assessment studies is well entrenched in international law. The Convention on Environmental Impact Assessment in a Transboundary Context (Espoo Convention) states that the Parties 'shall take the necessary legal, administrative or other measures to implement the provisions of this Convention, including, [...] the establishment of an environmental impact assessment procedure that permits public participation [...]'.[115] The Aarhus Convention implementation guide explains further the relationship between the obligation to conduct impact assessment studies and public participation[116]:

> the Aarhus Convention does not require an environmental assessment to be carried out. [...] However, if an environmental assessment is carried out (either EIA or SEA) then the public participation provisions of the Convention will apply. [...] Thus, one can conclude that [...] *public participation is in fact a mandatory part of environmental assessment*

In short, once (and if) the process of Impact Assessments will be assimilated within the SDM, it is expected that at least some level of public participation will find its way into this mechanism as well.

Public participation in the dispute settlement process Another way to increase public participation in SDM projects is via the investor-state dispute settlement system (ISDS). This dispute settlement system is reviewed elsewhere in this study (see chapter on investment law), and only a few brief comments will be made here. Just to remind the reader, it is argued that investment law mechanisms should be integrated into environmental law HCMs, including the ISDS which is provided by most investment treaties.[117] The public's participation is important not only during the project's establishment phase, but also throughout the project's life, including where disputes have arisen.[118] Importantly, the public can inform the arbitral tribunals on the impact that these disputes are having on human rights, including on the investors' behaviour in this context.

These public interventions could be useful for several reasons. For example, they can influence the arbitrators' understanding and interpretation of the law, including the role of complementary legal systems such as human rights treaties, as prescribed by Art 31(3)(c) of the Vienna Convention on the

115 See Article 2(2) of the Convention on Environmental Impact Assessment in a Transboundary Context (Espoo Convention).

116 United Nations Economic Commission for Europe, *The Aarhus Convention: An Implementation Guide* (2nd ed., United Nations 2014) <https://www.unece.org/fileadmin/DAM/env/pp/Publications/Aarhus_Implementation_Guide_interactive_eng.pdf> 122.

117 See a review of public participation mechanisms in the investment regime in de Brabandere, Gazzini and Kent (n 106).

118 See a full review in de Brabandere, Gazzini and Kent (n 106).

Law of Treaties.[119] Moreover, as recommended in the chapter on investment law, more recent investment treaties include references to Corporate Social Responsibility (CSR) codes and standards. Where such references do exist, arbitrators can use the information provided by the public and apply it in light of these CSR-related instructions. For example, where it is claimed that the investor has violated local communities' human rights, this information may influence the level of protection that could be claimed via investment law, *inter alia*, through the interpretation of definitions such as 'investment' or 'legitimate expectations'.

ITEM 3: EFFECTIVE REMEDIES

Item 3 in Knox's list addresses the provision of effective remedies for human rights' violations. Effective remedies for human rights violations are provided mostly where states are politically willing to provide such remedies, whether through their own domestic legal systems or by acceding to human rights conventions. In certain high-profile cases, local communities were left with no access to effective remedies,[120] whether through their home-state's courts or the courts of other nations.[121] These cases have attracted international attention, leading to international reforms: *inter alia* the conclusion of the UNGP on Business and Human Rights.[122]

Effective remedies through national courts The search for effective remedies in cases of human rights violations begins with domestic courts. Domestic courts may offer a variety of remedies (depending on the laws of each national legal system) and are without a doubt the most suitable forum for addressing such disputes. This legal route, however, could be restricted, especially where judicial systems that are unable (or unwilling) to rule against

119 Art 31(3)(c) instructs that treaties will be interpreted in light of 'any relevant rules of international law applicable in the relations between the parties'. As most of the international community has ratified at least some of the major human rights treaties (e.g. more than 170 states have ratified the International Covenant on Civil and Political Rights and the International Covenant Economic Social and Cultural Rights), investment arbitrators are obliged to consider these in the process of investment treaty arbitration.

120 See for example the case of the Ogoni People and their continuous search for remedy with respect to the responsibility of Shell vis-à-vis the events that took place in Nigeria during the 1990s, including the execution of nine persons and the violation of human rights.

121 See for example the high-profile Kiobel case, in which the US Supreme Court closed the possibility of relying the US Alien Tort Statute (ATS) for the purpose of litigating human rights violations abroad. See a review in Tyler Giannini and Susan Farbstein 'Corporate accountability in conflict zones: How Kiobel undermines the Nuremberg legacy and modern human rights' (2010) 52 *Harvard International Law Journal* 119.

122 The introduction to the UNGP states that the 'Guiding Principles are grounded in recognition of: [...] The need for rights and obligations to be matched to appropriate and effective remedies when breached'. United Nations guiding principles on business and human rights (2011) 1 (under 'General Principles'). See also Principle 22 of the UNGP ('remediation') and mostly Part III ('access to remedy').

the will of the executive. In such cases, claimants will have to search for a remedy elsewhere, whether in foreign courts or through alternative international legal mechanisms.

On some occasions, attempts were made to seek remedies in the courts of other nations. Foreign courts are far from an ideal choice of forum: these courts are remote from the events at hand and are likely to decline jurisdiction for a number of reasons (e.g. Forum Non Conveniens).[123] There could be however, reasons favouring an attempt to litigate before these courts. To begin with, foreign courts may be able to provide more effective remedies and relatively easy enforcement. Second, litigation in the investor's home country may also create public pressure on the investor, incentivising him to act in order to remedy the situation. Lastly, as discussed above, foreign courts may simply be the claimants' last and only hope.

Claimants approaching foreign courts are often hoping that these courts will find the legal grounding – as well as the political will – to provide extra-territorial rulings on events that took place outside of their jurisdiction. This option is far from certain: in the US for example, the Supreme Court's *Kiobel* ruling has *de facto* rejected the notion of extra-territoriality in cases where the events did not take place in this country.[124] This ruling effectively eliminated what was considered at the time as the main legal route for addressing US investors' human rights violations abroad.[125]

123 For a review see Jones Day 'Recent developments in suits against governments and corporations doing business abroad' (2013) <https://www.jonesday.com/en/insights/2013/05/recent-developments-in-suits-against-foreign-governments-and-corporations-doing-business-abroad>.

124 *Kiobel v. Royal Dutch Petroleum Co.* (2013) 569 U.S. 108. The Supreme Court stated in this case that

> On these facts, all the relevant conduct took place outside the United States. And even where the claims touch and concern the territory of the United States, they must do so with sufficient force to displace the presumption against extraterritorial application.
>
> See page 14 of this judgement.

125 Dodge mentions several other obstacles that will stand in the way of future attempts to bring human rights claims before US Courts. These include the fact that violations are likely to be the responsibility of daughter companies, rather than the 'targeted' US corporate; the fact that the law will not apply to foreign corporation and will concern only with certain customary international law-based violations; the reliance of some courts on the location of the violation as a decisive factor, excluding any possibility of extra-judicial jurisdiction. See more in William Dodge 'Introductory remarks by William S. Dodge' (2019) 113 *ASIL Proceedings* 159, 160–161. At the same time, it should be noted that certain authors believe that other routes may still be available to claimants in such cases See for example Roger Alford 'The future of human rights litigation after Kiobel' (2014) 89 Notre Dame Law Review 1749; Roxanna Altholz 'Chronicle of a death foretold: The future of U.S. human rights litigation Post-Kiobel' (2014) 102 *California Law Review* 1495.

In Europe, the approach displayed by domestic courts is described as 'mixed'.[126] On the most liberal end of the spectrum, the UK's Supreme Court ruled in *Vedanta vs Lungowe and others*[127] that UK-based parent companies could potentially owe a duty of care towards local communities in Africa, where these are affected by the operation of their subsidiaries. A similar approach (albeit with different reasoning) was presented by the Canadian Supreme Court. This Court ruled in *Nevsun vs Araya* that the violation of Customary International Law (notably human rights violations) could justify a claim for a remedy in Canadian courts.[128] Whether the British and Canadian courts established a trend remains to be seen: these two decisions are fairly recent[129] and their full impact is not yet known.[130] At the time of writing, it seems that the chances of human rights-related claims in foreign courts are uncertain and depend mostly on the jurisdiction in which they are attempted. Such attempts are also costly for claimants, and are often dependent on whether third parties are willing to promote cases at their own expense.

The 'Zero Draft' Treaty ('Legally Binding Instrument to Regulate, in International Human Rights Law, the Activities of Transnational Corporations and Other Business Enterprises')

Many of the obstacles discussed above could be remedied through the adoption of the currently negotiated 'Zero Draft' Treaty ('Legally Binding Instrument to Regulate, in International Human Rights Law, the Activities of Transnational Corporations and Other Business Enterprises' ('Zero Draft')).[131] If accepted, this treaty will revolutionise this area of law and overcome many of the obstacles currently standing in the way of victims seeking a remedy.[132] Importantly, it would allow victims to bring claims also in 'the Court of the State where the natural or legal person or association of natural or legal persons alleged to have committed the acts or omissions are domiciled'.[133] The Zero Draft further removes hurdles with respect to limitations, it provides a long list of rights for victims, and imposes a strong duty of prevention and due-diligence on states.

126 See review of European courts on this matter in Lucas Roorda 'Jurisdiction in foreign direct liability in Europe' (2019) 113 *ASIL Proceedings* 161.

127 *Vedanta Resources PLC and anor. v Lungowe and others* [2019] UKSC 20.

128 *Nevsun Resources Ltd v Araya* [2020] SCC 5.

129 The *Vedanta* ruling is from 2019 and the *Nevsun* ruling is from 2020.

130 One interesting development from the UK is the *Okpabi and others v Royal Dutch Shell Plc and Another* (Judgement, 12 February 2021), in which the UK Supreme Court effectively followed the line presented in the Vedanta Judgement.

131 The negotiated draft is available here: <https://www.business-humanrights.org/sites/default/files/documents/DraftLBI.pdf>.

132 See a review of this draft in David Birchall 'Between apology and utopia: the indeterminacies of the zero draft treaty on business and human rights' (2019) 40(2) *Suffolk Transnational Law Review* 37.

133 See Article 5 of the Zero Draft (n 131).

The reader should be aware that this is not the first attempt to negotiate such a treaty,[134] and it is questionable whether the negotiations will, eventually, lead to any result. The Draft itself is not free from problems; different elements in it have been heavily criticised,[135] including by John Ruggie himself.[136] A detailed analysis of the Zero Draft will not be provided in this study. It is nevertheless important to mention this Draft; the (rather slim) possibility that it will, eventually, be adopted, and the impact that such a development will have on the accessibility of remedies in cases of human rights violations.

Effective remedies through international institutions' grievance mechanisms

The search for an effective remedy need not end with domestic courts, which may, or may not, offer a remedy. A major criticism on the CDM was its lack of a robust grievance mechanism, through which affected individuals and communities will be able to raise issues concerning the impact of CDM projects on their enjoyment of human rights.[137] According to Olawuyi, the lack of grievance mechanisms 'has entrenched a culture of "approvals" that excludes the common citizen, whose fundamental human rights may be repressed by the home state and that repression endorsed by the UNFCCC'.[138]

Knox himself mentions certain useful models based on which the SDM could base its own grievance mechanism. The first model relies on two examples: Standard 8 of the Global Environment Facility (GEF) Policy on Agency Minimum Standards on Environmental and Social Safeguards,[139] and the grievance mechanism arrangements that are mentioned in the Adaptation Fund's Environmental and Social Policy.[140]

These two examples do not represent 'classic' grievance mechanisms. The GEF and the Adaptation Fund both require their *partner agencies/*

134 Birchall (n 132).

135 Doug Cassel 'At last: A draft UN treaty on business and human rights' (2018) *Letters Blogatory* <https://lettersblogatory.com/2018/08/02/at-last-a-draft-un-treaty-on-business-and-human-rights/#more-27105>; Carlos Lopez 'Towards an international convention on business and human rights' (2018) *OpinioJuris* <http://opiniojuris.org/2018/07/23/towards-an-international-convention-on-business-and-human-rights-part-i/>.

136 John R. Ruggie 'Comments on the "zero draft" treaty on business & human rights' (2018) *Business & Human Rights Resource Centre* <https://www.business-humanrights.org/en/comments-on-the-%E2%80%9Czero-draft%E2%80%9D-treaty-on-business-human-rights>.

137 Olawuyi (n 63) 137.

138 Olawuyi (n 63) 137.

139 GEF, *Policy on Agency Minimum Standards on Environmental and Social Safeguards* (2011) GEF/C.41/10/Rev.1 <https://www.thegef.org/sites/default/files/council-meeting-documents/C.41.10.Rev_1.Policy_on_Environmental_and_Social_Safeguards.Final%20of%20Nov%2018.pdf>.

140 See para 34, Adaptation Fund, *Environmental and Social Policy* (2013) <https://www.adaptation-fund.org/wp-content/uploads/2015/09/Environmental-Social-Policy-approved-Nov2013.pdf>.

implementing entities (e.g. regional development banks) to maintain effective grievance mechanisms, rather than manage the grievance mechanisms on their own.[141] Importantly, this requirement is complemented with an additional mechanism – a contact point within these funds[142] – through which affected persons/communities will be able to contact these institutions directly, and bypass partners' agencies/implementing entities' grievance mechanisms. The presence of this supplementary element is crucial as it allows a certain control over the handling of these complaints by entities such as development banks. This completing element also ensures that some responsibility (and liability) will remain with these funds, preventing the complete 'outsourcing' of the obligation to provide effective remedies.

The role of the GEF's contact ('Conflict Resolution Commissioner', or the 'Commissioner') is especially interesting in this context. The Commissioner is instructed to take active steps to 'resolve the dispute'. He/she is instructed to work directly with the complainant,[143] as well as 'with member countries and GEF Agencies to help resolve disputes and address complaints'.[144] This function implies a certain conciliatory role.[145] As discussed in the chapter on insurance HCMs, the existence of a powerful World Bank-backed conciliator may ensure that states take complaints seriously and that these conciliation attempts will not be ignored.

A more recent example involves the GCF's Independent Redress Mechanism (IRM).[146] The IRM is authorised to address complaints, or to initiate proceedings on its own, concerning the operation of GCF-funded projects.[147] The IRM is further authorised to monitor the implementation of its recommendations on past disputes.[148] In addition to the IRM, the GCF's Environmental and Social Policy also requires that 'accredited entities establish activity-specific grievance redress mechanisms as appropriate'.[149] Some may label this requirement to add an additional layer as redundant

141 See Standard 8, GEF *Policy on Agency Minimum Standards on Environmental and Social Safeguards* (n 139) 29.

142 The GEF designated a Conflict Resolution Commissioner for this purpose, while the Adaptation Fund simply as affected persons/communities to contact its secretariat.

143 Para 11, GEF *Policy on Agency Minimum Standards on Environmental and Social Safeguards* (n 139).

144 See Annex III: Background on GEF Conflict Resolution Function, GEF *Policy on Agency Minimum Standards on Environmental and Social Safeguards* (n 139).

145 Indeed, the process, as described on the Commissioner's webpage, reminds conciliation/mediation, see <https://www.thegef.org/conflict-resolution>.

146 Green Climate Fund, *Procedures and Guidelines of the Independent Redress Mechanism* (GCF 2019) <https://www.greenclimate.fund/sites/default/files/document/procedures-guidelines-irm.pdf>.

147 Green Climate Fund, *Procedures and Guidelines of the Independent Redress Mechanism* (n 146) para 2(b) and 2(c).

148 Green Climate Fund, *Procedures and Guidelines of the Independent Redress Mechanism* (n 146) para 2(b) and 2(d).

149 Green Climate Fund, *Environmental and Social Policy* (GCF 2019), para 12(c).

or even as an overkill. There are, however, several obvious advantages: an additional 'specific' mechanism will operate closer to the communities and thus, potentially be more approachable (especially where the alternative – the IRM – is based on continents away from the events). Such a specific mechanism will be familiar with the projects' specifics, history, dynamics and events, and its permanent present may, ideally, be used as a facilitating power. The GCF further ensures that the establishment of 'specific' mechanisms will not replace, or limit access to its own IRM. Its Environmental and Social Policy states in this respect[150]:

> GCF recognizes that local or project level grievance mechanism can provide an effective and direct remedy to complainants, and encourages the use of such mechanisms whenever possible. This recognition does not limit in any way the ability of complainants to access the GCF independent Redress Mechanism directly.

Knox himself advocates for a more ambitious model. The SDM, he argues, should establish its own grievance mechanism, enabling it to address complaints directly.[151] As examples, he proposes that the SDM's grievance mechanism will be based on the World Bank's Compliance Advisor/Ombudsman (CAO). The CAO acts as an 'independent accountability mechanism'[152] for both the International Finance Corporation (IFC) and MIGA. Its role is *inter alia* to investigate complaints made by external parties affected by MIGA or IFC's projects in collaboration with all affected parties, 'following a flexible process aimed primarily at correcting project failures and achieving better results on the ground'.[153] Many more models can be found elsewhere, including those adopted by development banks. The Inter-American Development Bank (IDB), for example, has an Independent Consultation and Investigation Mechanism (MICI)[154] whose role is to investigate allegations of IDB-funded projects' violation of IDB's policies on issues such as environmental protection,[155] involuntary resettlement,[156] gender equality[157] and indigenous

150 Green Climate Fund, *Environmental and Social Policy* (n 149), para 76.

151 Knox (n 77) 11–12.

152 See CAO's website <http://www.cao-ombudsman.org/>.

153 Office of the Compliance Advisor/Ombudsman (CAO), Terms of reference <http://www.cao-ombudsman.org/about/whoweare/documents/TOR_CAO.pdf> 2.

154 See more on the process on MICI's website <https://www.iadb.org/en/mici/our-mandate?open_accordion=1>.

155 Inter-American Development Bank, *Environment and Safeguards Compliance Policy* (2006) <http://idbdocs.iadb.org/wsdocs/getdocument.aspx?docnum=665902>.

156 Inter-American Development Bank, *Involuntary Resettlement* <https://www.iadb.org/en/about-us/about-us-7>.

157 Inter-American Development Bank, *Operational Policy on Gender Equality in Development* (2010) <http://idbdocs.iadb.org/wsdocs/getdocument.aspx?docnum=35428399>.

people.[158] The European Bank for Reconstruction and Development has its own Project Complains Mechanism (PCM).[159] Importantly, in order to make this process more approachable, the European Bank provides a template complaint form, and a designated PCM officer for assistance with the submission procedure. The PCM roles includes a compliance review function, through which it ensures the funded projects are complying with the Bank's policies (including environmental policy). Another important element is the Bank's Problem Solving Initiative, whose role 'might include independent fact-finding, mediation, conciliation, dialogue facilitation, investigation or reporting'.[160] In short, there are many other examples of grievance mechanisms in international institutions[161] that the negotiators of the future SDM could learn from.[162]

The architecture of any future arrangement might dictate the exact form of the appropriate grievance mechanism. In the case of a bilateral arrangement between two states only, this study has recommended the establishment of certain administrative bodies that will *inter alia* act as 'focal points' for investment promotion activities, coordinate trade-related benefits (e.g. pre-approving project-based EGS and Intellectual Property (IP) rights) and more. These administrative units could also be entrusted with the operation of grievance mechanisms. The process that these administrative units will adopt could be based on any of the examples set above: for example, include an early 'facilitative', informal approach, as applied by GEF's Conflict Resolution Commissioner, followed up by a formal investigation, including the option to sanction investors (e.g. excluding these from future HCMs). Another possibility is simply to rely on existing institutions' own mechanisms. Notably, as proposed by Knox, the new SDM could rely on the CAO's services, given that this institution already possesses the expertise and resources to pursue complaints of this nature.

In addition to a grievance mechanism, and in line with the Zero Draft approach, the Parties to such agreements could decide to grant the courts

158 Inter-American Development Bank, *Operational Policy on Indigenous Peoples and Strategy for Indigenous Development* (2006) <http://idbdocs.iadb.org/wsdocs/getdocument.aspx?docnum=2032081>

159 European Bank for Reconstruction and Development 'Project complains mechanism' <https://www.ebrd.com/work-with-us/project-finance/project-complaint-mechanism.html>.

160 For more information, policies and procedural operation rules of this mechanism, see European Bank for Reconstruction and Development 'About the project complains mechanism' <https://www.ebrd.com/work-with-us/project-finance/project-complaint-mechanism/about.html>.

161 See a list of mechanisms with links to relevant website <https://www.thegef.org/sites/default/files/documents/agency_mechanisms_conflict_resolution_accountability_20190321.pdf>.

162 An extensive review of these models is available in the literature. See for example Justine Nolan, Michael Posner, and Sarah Labowitz 'Beyond Kiobel: Alternative remedies for sustained human rights protection' (2013) 107 *AJIL Unbound* 48.

of investors' home countries specific, extra-judicial jurisdiction to litigate human rights violation abroad. This special authorisation could explicitly target only projects that were approved by the administrative unit discussed above, or more widely, the operation of all national firms abroad (e.g. where courts have already ruled that they have such jurisdiction). Adding this additional layer is important, especially as the ability of grievance mechanisms to impose remedies such as compensation is often limited.

ITEMS 4 AND 5: ADEQUATE SAFEGUARDS

Item 4 on Knox's list refers to 'protection against environmental and social risks'. Item 5 is in a way a refinement of item 4, and it refers to 'the protection of the most vulnerable'. These two items are addressed here jointly, as they are in essence addressing the need to impose safeguards on investors' activity in order to ensure PCD and coherence between the environmental and human rights spheres.

As an example, Knox refers to the Adaptation Fund Environmental and Social Policy.[163] As already discussed above, this Policy is useful mostly as it imposes obligations on intermediaries, such as the Fund's implementing entities (e.g. regional development banks) which operate more directly with the investors. The Policy requires these implementing entities to[164]:

> (i) have an environmental and social management system that ensures environmental and social risks are identified and assessed at the earliest possible stage of project/programme design, (ii) adopt measures to avoid or where avoidance is impossible to minimize or mitigate those risks during implementation, and (iii) monitor and report on the status of those measures during and at the end of implementation.

The Adaptation Fund Environmental and Social Policy also *directly* requires funded projects to comply with a list of principles, *inter alia*, concerning the protection of marginalised and vulnerable groups,[165] gender equity,[166] indigenous peoples[167] and more.

A closer look at the operation of 'implementing entities' reveals additional, slightly more complex models. One example of significant importance is the Asian Development Bank's (ADB) Future Carbon Fund (FCF).

163 Adaptation Fund, *Environmental and Social Policy* (Approved in November 2013) <https://www.adaptation-fund.org/wp-content/uploads/2015/09/Environmental-Social-Policy-approved-Nov2013.pdf>.

164 Adaptation Fund, *Environmental and Social Policy* (n 163) para 10. More details instructions on each of these requirements are available in the Policy itself.

165 Adaptation Fund, *Environmental and Social Policy* (n 163) para 14.

166 Adaptation Fund, *Environmental and Social Policy* (n 163) para 16.

167 Adaptation Fund, *Environmental and Social Policy* (n 163) para 18.

The FCF's role is to support and finance CDM projects by pre-purchasing CER units that are expected to be generated by these projects, for an agreed fixed price.[168] The FCF is funded by a list of donors, including (but not only) governments, and as such could be considered as an HCM. In the context of this chapter, the operation of the FCF is interesting due to its self-imposed Sustainable Development standards. The ADB has developed a list of 41 indicators, ranging from air-quality to employment quality, enhanced health services and women empowerment.[169]

The European Bank for Reconstruction and Development's (EBRD) Environmental and Social Policy is another example in which special attention to PCD could be found. With respect to human rights, the EBRD's policy states[170]:

> The EBRD is committed to the respect for human rights in projects financed by EBRD. EBRD will require clients, in their business activities, to respect human rights, avoid infringement on the human rights of others, and address adverse human rights risks and impacts caused by the business activities of clients.

The EBRD Policy includes a list of specific Performance Requirements (PRs) including on environmental matters such as pollution and biodiversity. Importantly for this section, this Policy also covers social matters *inter alia* labour and working conditions, land acquisition, restrictions on land use and involuntary resettlement, indigenous peoples and cultural heritage. The Policy further requires that

> [a]ll projects undergo environmental and social appraisal both to help EBRD decide if the project should be financed and, if so, the way in which environmental and social risks and impacts should be addressed in planning, implementing and operating a project.[171]

A different set of self-imposed guidelines was adopted in 2018 by the GCF.[172] The GCF's Environmental and Social Policy is aimed at managing

168 ADB, *Future Carbon Fund: Delivering Co-benefits for Sustainable Development* (ADB 2017) <https://www.adb.org/sites/default/files/publication/389821/future-carbon-fund.pdf> 6–7.

169 ADB (n 168) 10–11.

170 European Bank for Reconstruction and Development, *Policy Environmental and Social Policy* (2019) <file://ueahome/eresssf4/sdtl4cku/data/Downloads/ebrd-environmental-and-social-policy-2019.pdf> para 2.4.

171 EBRD (n 170) para 4.7.

172 GCF, *Decision B.19/10* (2018) <https://www.greenclimate.fund/sites/default/files/document/gcf-b19-43.pdf>.

the social and environmental risks that may be created by its financed projects. Financed projects are required to[173]:

> Avoid, and where avoidance is impossible, mitigate adverse impacts to people and the environment;
> Enhance equitable access to development benefits; and
> Give due consideration to vulnerable and marginalised populations, groups, and individuals, local communities, indigenous peoples, and other marginalized groups of people and individuals that are affected or potentially affected by GCF-financed activities.

The GCF's Environmental and Social Policy further provides a list of 18 'guiding principles' for the implementation of these goals. Most of these principles are fairly vague, and are addressing a variety of issues ranging from human rights to integration, gender equality and labour standards.[174] Some of these principles are addressing more specifically environmental protection: Principle (a), for example, reflects the principle of integration: a requirement to integrated environmental protection into processes such as the accreditation and monitoring of investments. Another relevant example is found in Principle (r): an obligation to design and implement financed activities 'in a manner that will protect and conserve biodiversity and critical habitats, ensure environmental flows of water, maintain the benefits of ecosystem services, and promote the sustainable use and management of living natural resources'.[175]

B Can soft-law mechanisms be the answer?

Another way to integrate human rights into the SDM (or any international law HCM) is through reliance on soft law. In an ideal world, Knox's 'check list' would be considered by states, be incorporated as 'hard laws' into the SDM and ensure PCD through the adoption of sufficient human rights safeguards. If history provides any indication, however, it is likely that many of the items on Knox's 'check list' will not be followed in full, if at all.

One, perhaps more feasible way, to incorporate these elements into the SDM (or any other international RE-related HCM) is the use of soft-law mechanisms. The advantages of soft-law mechanisms are clear. To begin with, these tools are more politically feasible. As reviewed above, some of the CDM's problems originated from the lack of political agreement on terms such as 'sustainable development', or on clear indicators that will

173 *Decision B.19/10* (n 172) Art 3(1)(3).
174 *Decision B.19/10* (n 172) Part IV.
175 *Decision B.19/10* (n 172) Part IV (a).

confirm investments' sustainability. Soft-law mechanisms could, as their name implies, soften resistance and lead to a wider acceptance of standards.

Soft-law mechanisms could also be developed from the bottom-up, via non-governmental institutions, and away from the neutering nature of political compromise. Even where these guidelines are rejected by governments, they can still garner sufficient public support and appeal to investors and funders. These, in turn, may choose to follow them unilaterally, on top of (and regardless of) existing, less demanding binding legal frameworks.

Specific soft-law guidelines are useful for other reasons as well. In light of increased international scrutiny over the activity of foreign investors (and indeed also the rise in CSR-based litigation), soft-law guidelines could provide investors with a useful preventative 'defence'. Investors that will follow to the letter, say, the UNGP on Business and Human Rights, may be better protected against criticism (or legal liability) once future disputes arise.

Another relevant point can be found in an ADB report, which stresses the importance of soft-law mechanisms in terms of providing clearer guidance to investors.[176] This report explains that while investors may be aware of the need to increase their projects' sustainability, they do not always have the tools to systematically evaluate their projects' sustainability.[177] Soft-law solutions may come in handy in this respect, by providing investors with detailed standards, criteria and best practices.

Importantly, soft-law guidelines could also improve the SDM's alignment with the SDGs. As briefly discussed above, one of the main criticisms concerning the operation of the CDM was its questionable link with sustainable development. The KP did not provide any measurable standards for assessing what the contribution of CDM projects to 'sustainable development' should be, and host states were allowed to define these 'contributions' on their own. Moreover, once a CDM project has been approved, there is no obligation to monitor this project's operation and ensure that indeed it is delivering on any SD benefits promised at the approval stage. A clear set of soft-law-based guidelines could address this problem and provide investors with an indication as to their compliance with the SDGs. Importantly, such guidelines will be beneficial also for the purpose of compliance: while non-binding in nature, NGOs could use these as a benchmark for their scrutiny of big investment projects, shaming and naming relevant actors.

A number of authors have indeed proposed to contend with these issues through soft-law mechanisms. Recognising the difficulties in adopting a unified, top-down, international SD criteria. Olsen et al. have advocated for a voluntary Sustainable Development SD labelling tool.[178] These authors are promoting a ranking-based model, which relies on formulas

176 ADB (n 168) 22.
177 ADB (n 168) 22.
178 Olsen et al. (n 27).

that quantify a variety of SD-related elements. This measuring tool is supposed to produce a unified 'SD score' for each project.[179] The advantages of such a labelling tool, they explain, are significant: it will create unified standards; reduce transaction costs; and, importantly, establish a 'premium' market for labelled, measured and monitored emission units, which will be appealing to some buyers, especially in light of the CDM's murky reputation.[180]

Another initiative that makes use of soft-law certification is the Gold Standard for the Global Goals ('Gold Standard'). The Gold Standard is an independent initiative led by NGOs such as the World Wide Fund for Nature (WWF). The Gold Standard quantifies, and verifies, projects' activity in light of the SDGs. In 2017 an RE-specialised Standard was launched – the Renewable Energy Product Label. In order to get certified, a project must demonstrate *inter alia* a genuine, high-quality, additional RE contribution to the grid. Other elements that will have to be demonstrated are related to inclusivity and public participation and contribution to at least three SDGs (two of them must be SDG 7 and 13). Relevant projects are audited by an approved Gold Standard auditor and are required to apply a list of safeguards.[181]

Other relevant models exist as well,[182] and their adoption by investors is not uncommon.[183] As we witness a rise in the utilisation of soft-law benchmarks also in other international HCMs (notably investment treaties[184] and insurance mechanisms[185]), it will not be surprising to find in the negotiated SDM also references to soft-law mechanisms that will secure investments' impact on human rights and sustainability more broadly. It seems equally evident that states should at least consider the inclusion of soft-law arrangements in the negotiations of bilateral and regional arrangements that are addressing investment in the RE sector.

179 Olsen et al. (n 27).

180 Olsen et al. (n 27) 247.

181 Gold Standard for the global goals renewable energy label product requirement <https://globalgoals.goldstandard.org/wp-content/uploads/2017/08/500-GS4GG-Renewable-Energy-Label-Product-Requirements-v1.1.pdf>.

182 Axel Michaelowa, Aglaja Espelage and Stephan Hoch (n 28) 58; Christof Arens, Florian Mersmann, Christiane Beuermann, Frederic Rudolph, Karen Holm Olsen and Jørgen Villy Fenhann, *Mapping the Indicators. An Analysis of Sustainable Development Requirements of Selected Market Mechanisms and Multilateral Institutions* (German Emissions Trading Authority 2015).

183 Axel Michaelowa, Aglaja Espelage and Stephan Hoch (n 28) 58.

184 See review in Avidan Kent and Markus Gehring 'Investment law and the environment: Evolving international practice and norms' in Erika Techera, Jade Lindley, Karen Scott and Anastasia Telesetsky (eds.) *Routledge Handbook of International Environmental Law* (2nd ed. Routledge 2020), chapter 27.

185 See Chapter 4 in this study.

IV International environmental law HCMs and PCD: a limited prospect

This chapter has provided a cursory review of the main current, and future, RE-related international environment law HCMs. It seems that important parts of this field are geared towards the promotion of RE-related investment, and that this emphasis is unlikely to change in light of the adoption of the nascent SDM. Nevertheless, while the engagement with international investment as an objective is clear and openly pursued, the effort to design supportive laws and policies are being made in isolation.

To begin with, by pushing towards a CDM-like mechanism, it is likely that the SDM will address only certain investment barriers, mostly (if not exclusively) financial risks. By ignoring other legal fields (e.g. trade and investment law), this mechanism is unlikely to contribute to the mitigation of other investment barriers, such as political risks and inadequate infrastructure (including legal infrastructure). Also, while the financial benefits generated through the CDM are without a doubt important, they are far less significant than other relevant financial components, notably the cost of finance. A link with other legal areas (e.g. insurance HCMs or directed ODA) would have been useful in mitigating these financial barriers and increasing the effectiveness of environmental law HCMs.

Approaching only one set of concerns – and accordingly pursuing only a limited notion of PCD – is highly inefficient. Such a limited approach does not unlock the full potential of the private sector; many investment barriers remain unaddressed, and as a result less investment is taking place. The lack of PCD is disappointing also because the UNFCCC could have offered a useful multilateral and holistic framework for the promotion of RE in the developing world; a task that it is indeed trying to achieve,[186] but doing so while ignoring important tools.

There are many possible explanations for the lack of PCD within the UNFCCC. These include the desire not to complicate the already complex climate negotiations, and the fear of some developing nations that the inclusion of trade-related matters in climate negotiations will end up with 'green protectionism'.[187] Another possible explanation is more social in nature and

186 This task is stated in the UNFCCC text itself: Art 3(5) states:

> The Parties should cooperate to promote a supportive and open international economic system that would lead to sustainable economic growth and development in all Parties, particularly developing country Parties, thus enabling them better to address the problems of climate change.

187 World Bank, *World Development Report: Development and Climate Change* (World Bank 2010) <https://openknowledge.worldbank.org/handle/10986/4387> 251.

involves the views and the perceptions of different epistemic communities. Koskenniemi has famously written in this respect[188]:

> To be doing "trade law" or "human rights law", or "environmental law" or "European law" – as the representatives of those projects repeatedly tell us – is not just to operate some technical rules but to participate in a culture, to share preferences and inclinations shared with colleagues and institutions who identify themselves with that "box".

Koskenniemi's words fit well with theories on epistemic communities. As Haas explains, these groups tend to share, *inter alia*, 'value-based' beliefs, perceptions of rationales and understanding of cause-effect relationships.[189] The entrenchment in one's own group's perception is dangerous, as it can lead to the passing of (too) quick judgement on issues, even where these are beyond one's area of expertise. The immediate (almost automatic) rejection of trade/investment tools by many in the 'environmental' camp is damaging in this respect. One striking example is found in the UN Special Rapporteur on Human Rights and the Environment's report on the issue of human rights obligations relating to the enjoyment of a safe, clean, healthy and sustainable environment. The Rapporteur – Professor David Boyd – is an accomplished author, but hardly an expert in the area of investment law. His very brief, and yet extreme recommendation on investment arbitration,[190] was indeed highly misguided or to the least misinformed,[191] and it was certainly unrepresentative of the views of most researchers in this area of law. Such an approach will not improve PCD and may end up damaging the very objective that Boyd and others are trying to promote.

At the same time, it is clear that not all is so bleak and some positive developments are taking place. To begin with, the shift from a CDM to a *Sustainable Development* Mechanism is a promising step. Notably, it implies a much wider consideration in line with the all-inclusive term 'sustainable development'. Whether this promise of holistic sustainability will, indeed, deliver remains to be seen. Critically, the interaction with appropriate

188 Martti Koskenniemi, *International Law: Between Fragmentation and Constitutionalism* (The Australian National University 27 November 2006) 4.

189 Peter M. Haas 'Introduction: Epistemic communities and international policy coordination' (1992) 46(1) *International Organization* 1, 13.

190 'States should withdraw consent to arbitration or negotiate a carve-out for climate actions as a safeguard to protect against these types of lawsuits' Boyd (n 65) 81.

191 Boyd claims that 'Corporations have used Investor-State dispute settlement mechanisms in investment treaties to file lawsuits seeking compensation for lost profits when climate policies are strengthened, creating regulatory chill'. The vast majority of these 'lawsuits' were made following the weakening of climate policies, not their 'strengthening' and the investment regime is used in these cases in order to strengthen stability and investment conditions. See more in the chapter on investment law HCMs, in this study.

safeguards – notably human rights-related (but not only) – is receiving increased attention.

This integration of human rights considerations into the process of foreign investment is reflecting a much wider trend: the rise of CSR. Developments such as the publication of the UNGP on Business and Human Rights and the OECD Guidelines for Multinational Enterprises – as well as the wide consensus that has been formed around these guidelines – are already changing investors' standards and behaviour. As discussed above, as part of this trend, a range of environmental and social policies and institutions have been set up by actors such as funders, insurers and, indeed, also investors. It could therefore be that even regardless of the SDM, the inclusion of human rights safeguards is happening as sufficient forces have already been set in motion.

This is not to say that Knox's framework – and its potential integration within the SDM – is not useful. The incorporation of this framework into the SDM (or indeed in bilateral/regional framework) will signify an important step forward and will improve the sustainability of investment in the RE sector. Importantly, it will offer a comprehensive and harmonised approach towards the integration of human rights safeguards into the process of investment in the RE sectors. It will, nevertheless, be only one step among many others that will have to take place.

Conclusions, lessons learnt and the way forward

This concluding chapter arrives after two years of writing. It will not re-tell the story already told throughout this book, and neither will it re-address specific issues that have already been discussed. Rather, it will provide certain 'big picture' observations about the lessons learnt, and, importantly, signpost the way forward. For more specific recommendations, the reader is invited to re-visit the book's chapters.

I Lessons learnt

This study has addressed a topic of great importance and a challenge of a great scale. The root causes of energy poverty and the obstacles that prevent its alleviation will not disappear on their own. They require active and aggressive action from a variety of actors from all sectors. Public policies will transpire in different shapes and forms: this study has focussed on *international law*, and the role that this instrument can fulfil in increasing RE-related investment in the Global South.

As stated in the introduction to this book, the issues at hand are extremely vast and this study has not addressed them all. As also explained in the introduction, providing a textbook-scale review was never the author's intention. It was intended to address a (comprehensive) list of relevant issues, while demonstrating several points, notably the importance of international law Home Country Measures (HCMs), and the poor state of policy coherence in existing international laws and policies. These two elements are relatively untouched in international law literature, and even more so (and far more importantly) in states' practice. This study evaluated both of these terms in tandem: it found that while international laws do provide a variety of RE-related HCMs, these measures are often limited in terms of scope and ambition. It further found that these measures are by and large designed in isolation, and with very little attempt to integrate them with other international legal regimes.

Treaty-negotiators addressing the challenge embedded in Sustainable Development Goal 7 (SDG 7) will have to be aware of the need to improve policy integration. While debates on the fragmentation of international law

DOI: 10.4324/9780429277207-9

are often described as academic and hypothetical, this study has demonstrated that this assumption is very far from being true. The successful integration of different HCMs could improve the effectiveness of international instruments and maximise the benefits that states could achieve for every publicly invested dollar, pound or euro. This is a matter *of practice*, in the most technical sense. It relates to the design of rules at numerous levels and requires policy-makers to roll up their sleeves, identify the specific areas in which different regimes interact, address these interactions and improve linkages. It is no longer enough to design laws on, say, international trade, by looking only at international trade. Decision-makers must understand the impact that other areas of law may have on any given negotiated topic, including the *positive impact* (rather than the oft-addressed *negative* impacts, as discussed below) on the effectiveness of other regimes – and address these as well.

In this respect, this study provides certain insights into the wider debate on the fragmentation of international law. Importantly, it demonstrates that the focus of this debate has been misguided. Authors have mostly focussed over the years on two issues: the possibility of conflicting international courts' decisions, and the possibility of conflicting instructions provided by different treaties.[1] However, the impact of the fragmentation (i.e. the evolution of isolated legal regimes) is far deeper, and much more severe. The effect of the fragmentation is found in a variety of technical areas, where a range of factors and instruments should have been holistically addressed together, in a mutually supportive manner, but due to the isolation imposed by the fragmentation, they were not. The impact of the admittedly rather rare cases of normative conflicts is dwarfed in comparison to the much more common loss of potential linkages, compatibilities and mutually supportive laws and policies.

Further, this study sheds light on the lack of attention given to HCMs in the world of international law. Most of the HCMs discussed in this study are, in the least, under-developed, and while in some areas efforts are being made to improve this situation (e.g. environmental law or trade law), the current state of affairs is defined by limited ambition, a lack of specific attention to specific contexts and goals (as seen in the context of this study) and in some cases even regression (notably the field of investment law is increasingly becoming a *persona non grata* in many forums and states).

The utilisation of international law HCMs will have to improve. As reviewed throughout this study, developed nations have the resources,

1 See for example the work of Joost Pauwelyn 'Bridging fragmentation and unity: International Law as a universe of inter-connected islands' (2003–2004) 25 *Michigan Journal of International Law* 905. See also Bruno Simma 'Fragmentation in a positive light' (2003–2004) 25 *Michigan Journal of International Law* 905; Gilbert Guillaume 'Speech by His Excellency Judge Gilbert Guillaume, President of the International Court of Justice, to the Sixth Committee of the General Assembly of the United Nations' (27 October 2000); and great many others.

capacity and the moral responsibility to increase RE-related investment in the Global South. Even beyond these obvious rationales, improving RE-related international HCMs is entirely within the self-interest of developed nations. This is true for commercial reasons (markedly the existence of 1 billion potential, keen customers), but also in light of the alternatives. Notably, the dissemination of cheaper fossil fuel-based solutions in the Global South is still a viable possibility; one that will lock the planet into decades of pollution and doom the efforts to meet the 1.5°C target aspired to by the Paris Agreement.

The efforts to increase the quality and the quantity of international law HCMs are mostly in the hands of the Global North. Unlike other policies and commitments that are required from the developed world, this objective will present a genuine win-win scenario, as it is aimed at supporting *Global North industries*, leading to a line of subsidiary economic gains. Western governments understand the potential for providing green support packages at the domestic level: at the time of writing, unprecedented 'green deals' have been initiated in places such as the US, the European Union and the UK. It is time for the same logic to be applied to states' foreign policies, notably in the shape of international law HCMs.

II The challenge ahead

As discussed above, the challenge ahead is vast and complicated. But, on a more concrete level, it is useful to the discussion to make certain recommendations regarding the way forward.

A The venue and the form

First, there is the matter of the right venue and form. In other words, where should any reforms take place? And under which institution? And should the solution be multilateral, regional or bilateral in nature? All of these questions address the architecture of the solution, rather than the solution itself. These questions are nevertheless important as they will affect states' capabilities to reform and develop the international legal system.

Naturally, the most ideal solution to the situation discussed in this study is the conclusion of a global, specialised treaty that aims to promote RE-related investment in the Global South. In principle, such a mechanism is indeed being discussed: the United Nations Framework Convention on Climate Change (UNFCCC) Kyoto Protocol's Clean Development Mechanism (CDM) was designed to do just that, and the CDM's successor – the currently negotiated Sustainable Development Mechanism (SDM – based on Article 6 of the Paris Agreement) – should achieve similar objectives, albeit with slightly more attention to Policy Coherence for Development (PCD).

Notably, the work of John Knox – the previous UN Special Rapporteur on Human Rights and the Environment – has left a mark in this respect, and it seems reasonable to expect that the new SDM mechanism will be better attuned to human rights considerations. As explained in this study, other developments may indicate better alignment with environmental goals, but this possibility for the time being remains uncertain. Unfortunately, there is very little appetite to go further in terms of PCD and open the SDM negotiations to other forms of integration, including with most of the HCMs discussed throughout this study. While the shape and the scope of the negotiated SDM are currently unknown, it seems that the expected developments will hardly be ambitious enough and will not signify a meaningful change in approach.

This study does not ignore the importance of multilateral venues; institutions such as the WTO or the UNFCCC are extremely important, if only due to their wide coverage and global standard setting powers. Even partial advancements (e.g. securing human rights mechanisms in the negotiated SDM) will be extremely useful, even if other important instruments will eventually be ignored. At the same time, those with an appetite for development and reform will do better to look at the more modest regional and bilateral mechanisms. Indeed, the example of (mostly bilateral) development treaties presented us with the most advanced, PCD-oriented examples of international law HCMs. States will do well to follow this model.

Smaller arrangements (notably bilateral agreements) such as those discussed throughout this study are more likely to facilitate advancement and reform. Bilateral negotiations are much faster and much more flexible than their multilateral counterparts. Bilateral negotiations could include like-minded groups of developed and developing states, and address – holistically – a wide variety of issues, which may not traditionally sit within the same agreement. Bilateral agreements could be far more bespoke: for example, the necessity of specific investment protection rules could be evaluated in light of the relevant parties' history of respect for the rule of law and the efficiency of their local judicial systems.

Moreover, bilateral agreements could allow for relative flexibility in terms of future adjustments, or the inclusion of institutions that will determine future rights on a flexible, case-by-case manner. One proposal made in this study concerns the possibility of setting up bilateral joint committees that could approve trade liberalisation measures based on specific, proposed projects (rather than a pre-determined, inflexible Environmental Goods and Services (EGS) lists), as well as other relevant rights such as specific IP-related guarantees for specific projects. Further, these committees could have administrative responsibilities: for example, they could be designed as focal points, in which relevant information is collected and made available for investors (e.g. information on investment opportunities, as well as on the local legal environment). These committees could also administer other investment facilitation activities, as mentioned throughout this study (e.g. the provision of technical assistance, the administration of public insurance mechanisms) or even the administration of special funds. As recommended in Chapter 7 of this study, these 'focal points' could also administer grievance mechanisms (which will be based on environmental and human rights codes) and ensure the sustainability of supported investment.

A fairly limited model for these committees could be found in agreements such as the Trans Pacific Partnership (TPP) and the United States-Mexico-Canada Agreement (USMCA). The parties to the TPP established an Environment Committee, as well as Contact Points (Article 20.19) that are acting as a forum for discussion and are entrusted with the overseeing of the implementation of the TPP's environment chapter. Interestingly, under both the TPP and the USMCA, the relevant Environment Committee is also (rather vaguely) entrusted with the promotion of the early commitments made regarding the liberalisation of EGS. The TPP, for example, instructs[2]:

> 3. Accordingly, the Committee shall consider issues identified by a Party or Parties related to trade in environmental goods and services, including issues identified as potential non-tariff barriers to that trade. The Parties shall endeavour to address any potential barriers to trade in environmental goods and services that may be identified by a Party, including by working through the Committee and in conjunction with other relevant committees established under this Agreement, as appropriate.

This conclusion will not delve in to the variety of instruments that have been proposed throughout this study. Again, the point made here concerns the correct architecture that should be pursued. In this respect, the author proposes that the (largely bilateral) model of development agreements should be followed, and also to a large extent improved.

B Institutional reforms

As stated above, this study's main conclusion concerns the need to improve policy coherence. While this conclusion seems reasonable and even banal to a certain extent, achieving this goal remains extremely challenging. A key obstacle to this challenge concerns expertise. As the author has argued elsewhere,[3] the design of holistic and coherent policies requires a variety of experts from different areas. For example, trade experts know how to design trade laws but not how to align, and support where necessary, such laws with parallel policies from other areas. They lack the relevant expertise, and they miss the different perspectives and ethea that define and distinguish

2 Article 20.18(3) TPP.

3 Possible models include joint committees or working groups, or even the establishment of dedicated administrative divisions that will specialise in specific interactions. See Avidan Kent 'Implementing the principle of policy integration: Institutional interplay and the role of international organizations' (2014)14 *International Environmental Agreements* 203.

different epistemic communities. In this respect, Koskenniemi famously commented[4]:

> To be doing "trade law" or "human rights law", or "environmental law" or "European law" – as the representatives of those projects repeatedly tell us – is not just to operate some technical rules but to participate in a culture, to share preferences and inclinations shared with colleagues and institutions who identify themselves with that "box".

In the context of this study, it is clear that the design of international law HCMs will require experts in areas such as investment, trade, environment, insurance, climate, finance, human rights, taxation and more. This mixture of expertise will require a proper institutional setting. States will do well to think of appropriate institutional mechanisms through which different experts will be able to mix and collaborate in the process of policy design. The author of this study has proposed certain models in the past[5] and has no wish to recycle his previous work here. It will suffice to say that states will have to consider the right institutional setting for such negotiations and recognise the importance of plurality in this context.

C Shifting the focus and nature of obligations

The above point addresses the architecture of the required legal arrangement. The following point concerns the type of rules that should be addressed. In a nutshell, it seems that the issue of policy-linkages is addressed in international law mostly through what is defined in this study as 'defensive policies', namely policies that are focussed on ensuring host states' regulatory flexibility. When considering policy-linkages, states mostly see potential problems. They are concerned about the possibility of negotiated laws and policies negatively affecting the protection of other public interests (e.g. human rights, environmental protection), and therefore attempt to ensure certain regulatory flexibility in order to guarantee the protection of those important public interests.

The discussion regarding these defensive mechanisms (e.g. exceptions, carve-outs, codes that ensure investors behaviour) is without a doubt important. However, defensive policies only address a very limited part of the story of policy-linkages – those limited instances in which foreign investors' activities are in conflict with local communities environmental, social or human rights. These instances are without a doubt meaningful and should not be ignored. But let us remember that for every investor-community/state

4 Martti Koskenniemi 'International law: Between fragmentation and constitutionalism', Canberra (27 November 2006).

5 Kent (n 3).

dispute there are hundreds of interactions which lead to positive results (e.g. jobs, income, technology transfer), and in the vast majority of cases the parties do get along. The challenge of actively promoting RE-related investment in the Global South is evermore enormous in terms of scale, impact and importance. States will do well in this respect to invest more time, research and attention to the development of international law HCMs, rather than to the already well-explored defensive mechanisms. The focus, if to borrow from the world of sports, should shift from playing defensively to actively trying to score.

This final conclusion is relevant also in terms of academic research. A review of the literature will confirm that the discussion on defensive mechanisms has mostly been exhausted; mountains of dissertation, books and journal articles have already addressed the frankly limited, and simple, tools that are available for countries in this context. In terms of contribution to existing knowledge, researchers would do well to give this aspect a rest and explore the more challenging, and far wider (and admittedly far more technical) world of investment promotion measures, and other international law HCMs.

D And what role for academics?

The author has stated above that this topic is hardly 'academic'. But this is not to say that academics should not be involved; in fact, the opposite is true. Academics will have to support the development of solutions to the issues at hand. They can do so in a variety of ways, and, importantly, with different degrees of focus. This study has taken a 'bird's eye' approach, and it is strongly believed that this perspective is not found in enough academic studies today. Academics tend to specialise, to find one topic and dig as deeply as possible. Only too often this digging process is in one direction only, and ends only with the researcher's retirement party (and often thereafter). The result of this intensive specialisation is certainly useful: it creates the deepest and most intimate knowledge of a delimitated area. However, the flaws of this approach are also evident; the lack of capability to truly connect one's expertise with other, wider problems, and the entrenchment of the fragmentation.

This author will end this study with a call for wider perspectives, the study of linkages and the examination of different areas and worlds. The most significant and rewarding learning experience happens when one is pushed outside of one's comfort zone. Albert Einstein is known to have said that 'every fool can know. The point is to understand'. True understanding will only be achieved where wider perspectives are explored, and researchers will do well to remember this lesson.

Index

Note: Page numbers followed by "n" denote endnotes.

www.ingramcontent.com/pod-product-compliance
Lightning Source LLC
LaVergne TN
LVHW021123300926
849160LV00010B/832